S
T
WOLSZTYN

For the late Howard Jones MBE, founder of the Wolsztyn Experience

Crumps Barn Studio
Syde, Cheltenham GL53 9PN
www.crumpsbarnstudio.co.uk

Cover design by Lorna Gray
Map of the routes by Bob Bowden
All photographs © the author, unless otherwise credited

All our books are printed on responsibly sourced paper from managed woodlands and recycled material. Printed in the UK by CMP, Poole.

ISBN 978-1-915067-44-9

A PERSONAL STORY OF

THE WOLSZTYN EXPERIENCE

STEAMING THROUGH WOLSZTYN

JEFF NICHOLLS

The trains, the characters and five thousand miles on Polish footplates

Crumps Barn Studio

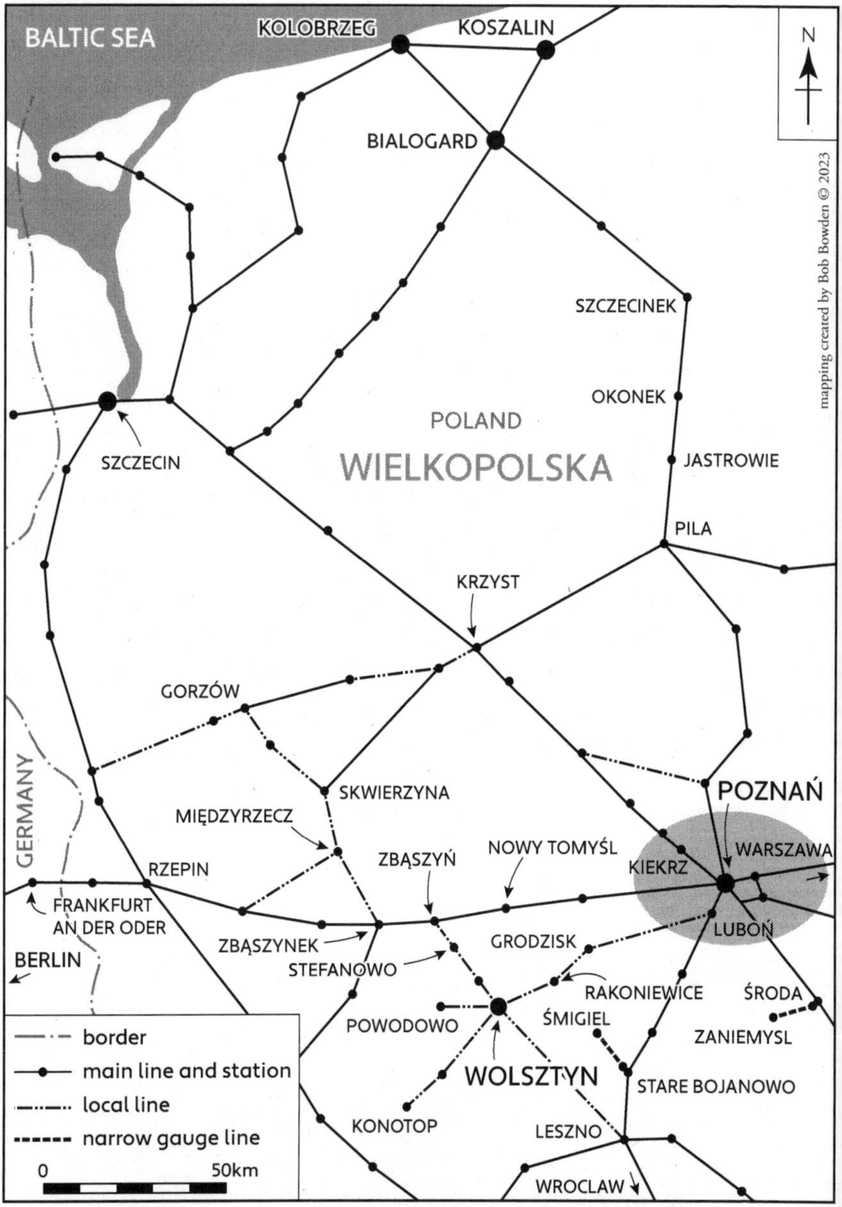

Schematic map of the railways in Western Poland
(for clarity, a number of local lines have been omitted)

INTRODUCTION

DZIEŃ DOBRY. In the west of Poland, just a few hours travel from the German capital of Berlin, lies the town of Wolsztyn. On the face of it, this is an ordinary Polish town: shops, offices, schools, factories, a hospital, cafes, restaurants and, Wolsztyn's most striking building, a Baroque-style church dating back to the late 1700s. One thing, however, distinguishes it from any other town in the world, a unique phenomenon in this age of efficient electric and diesel power – it has real working steam trains.

Steam locomotives haul regular timetabled passenger trains the whole year round on a state-owned railway system. Such sights have disappeared in every other corner of the world, but not yet from here. Incredibly, until very recently, anyone from anywhere in the world could – by paying – participate in the running of the steam-powered trains. Not only were they allowed to participate, but they were allowed to ride in the cabs of the steam locomotives. Not only were they allowed to ride in the cabs, but they were officially allowed to drive and fire the steam engine which hauled the trains! Surely this could not happen in our safety conscious world? It could. It did. For over a quarter of a century in this small but now world-famous Polish town, it was possible to work on everyday steam-hauled trains full of real, paying passengers. This is the story of my small part in those journeys.

The town of Wolsztyn and its steam trains have featured in several television programmes, notably being visited by

Michael Palin and the colourfully-dressed Michael Portillo in their entertaining travelogues. The Wolsztyn Experience itself was founded in 1997 by Howard Jones as a means of collaborating with Polish Railways to keep some steam going in one small corner of Poland so that it could be experienced by as many people as possible. However, the 'Experience' extended far beyond simply driving and firing steam locos; there was much, much more to it than that. There was the whole business of fending for yourself in a country where the language is very different from your own, and where the people have an outlook on life which differs in many ways as well. You would learn new skills, taught by diverse characters whose grasp of the English language varied from good to almost non-existent. You would probably learn more about yourself in the process as you would be expected to stand on your own two feet; nobody would 'hold your hand' through your time in Wolsztyn. For me, these experiences were part of the challenge, the excitement, the occasional drama, the frequent frustration but the sheer unadulterated pleasure and satisfaction which came from what I did.

If I set out for Wolsztyn in 2002 wearing rose-coloured glasses, they were quickly removed in a steep learning curve. Working on the footplate of a steam locomotive is far from glamorous; it is dirty, gritty and hard work – with an element of danger thrown in – but ultimately, it is very rewarding and very satisfying!

WOLSZTYN

The Society House in Wolsztyn

'Howard, there's a car in the house ...'

1

INTO THE GREAT UNKNOWN

GETTING THERE

My acquaintance with Wolsztyn and the wonderful country of Poland began in 2001 when lifelong friend Julian and I were debating how to celebrate our 50th birthdays. The Wolsztyn Experience was suggested; a decision made. Several phone calls later, travel arrangements were put in place by Howard's brother, Trevor.

Easter Sunday 2002, which fell on 1st April, found us at Stansted Airport bound for Berlin on the first leg of our Voyage into the Great Unknown. Even the journey to Wolsztyn was an adventure; we flew with Buzz Airlines, a short-lived, low-cost subsidiary of KLM. Before it even took off, our bright yellow plane was declared a failure and we had to wait several hours for a replacement. By his own admission, Julian is a dreadful flier, and his anxiety was not helped by the failure of our first plane, so he took advantage of the wait to down a few G & Ts to 'steady his nerves'. I stuck to water! After landing at the historic – but now defunct – Tempelhof airport, grabbing our bags from the carousel and hastening through Passport Control, we dashed outside to launch ourselves into one of Berlin's ubiquitous, hearing-aid-beige Mercedes taxis.

Fortunately, our connection in Berlin was generous.

Nevertheless, an anxious though swift taxi ride (*'Bahnhof Zoo schnell bitte'* – my best German) from the airport to Berlin's Zoo Station ensued. We were feeling rather warm after our exertions, but as we drove away from the airport, the driver began to close the taxi's windows.

'Finster offnen ist gut,' I said, summoning up more of my German.

Turning round to us in the back, the driver affirmed with pride, *'Zis Mercedes Benz has air conditioning.'*

And with that, he firmly closed the electric windows ... We passed remnants of the Berlin Wall but were in no mood to take in the sights, much less photograph them. We made Zoo Station with a quarter of an hour to spare.

Spot on time, our electric-hauled train glided in and we took our reserved seats. Because it was Easter Sunday, we found that we had virtually the whole train to ourselves. Our comfortable Berlin-Warsaw express passed through the centre of the city, the elevated tracks giving us excellent views of grand buildings like the Reichstag; we even thought we caught a fleeting glimpse of the Brandenburg Gate. Leaving the city behind, we headed into the flat, sandy, forested countryside of Eastern Germany. We had planned to grab something to eat in Berlin but fate had decreed otherwise so we decided to treat ourselves and visit the restaurant car, where we were outnumbered by the very attentive staff.

Our train took about an hour to reach the border town of Frankfurt Oder, stopping in the station there for the formalities of leaving Germany to be carried out. In those far-off days of 2002, Poland was two years away from joining the European Union, so it was a slightly anxious time when four

Polish Border Police – and their automatic rifles – boarded the train shortly before it pulled out of the station. No smiles, no exchanges of pleasantries, just the one-word command: 'PASSPORT!' No '*prosze*' – not that we then knew the Polish for 'please'. There followed what felt like an eternity of intense scrutiny; after all, we were amongst a mere half-dozen passengers so there was no hurry. Eventually, our documents were stamped – grudgingly we felt – as we crossed the broad River Oder, which forms the border between the two countries. 'O' Level History lessons came back to me as I remembered the Oder-Neisse line dividing the two countries. Pointing this out to my companion – and trainspotter pal from our school days in Wolverhampton – Julian pleaded ignorance and claimed, quite plausibly, not to have paid attention in history lessons!

Our adventure began in earnest as we passed into the unknown land of Poland, its flat, sandy terrain looking little different from eastern Germany. We soon saw our first Polish station, Słubice; shortly after that we stopped at Rzepin, where we noted a huge 2-10-0 loco plinthed on one of the platforms.

Returning from the restaurant car to our empty carriage, we found that Julian's leather jacket was missing. Amongst other things, it had contained his wallet. After a brief search of all the obvious places, we sought out the guard, who spoke little English but some German. Reluctantly, he accompanied us on a search of the train. Thankfully, the jacket was swiftly discovered in a toilet in the next carriage along from ours. The wallet was still in the jacket; some £80 in UK currency was not in the wallet. We came to the conclusion that the border guards had helped themselves. A lesson learned the hard way.

Some three hours after departing Berlin, along with two more passengers who were also participating in the 'Experience', we had to leave the comfort of the Berlin-Warsaw Express at a place with the unpronounceable name of Zbąszynek where, we had been assured, we would be met by someone who would take us on to Wolsztyn. And so we were; Bogdan spoke little English, we spoke no Polish. The word 'Wolsztyn' was mentioned, hands were shaken all round and, from that unpronounceable place, we were driven to our destination, a journey of some three quarters of an hour.

Our chauffeur was a lugubrious, chain-smoking, leather jacket-wearing man, sporting the droopy moustache which appears to be obligatory amongst his compatriots. Bogdan drove at high speeds, along pitch-dark country roads lined with warning signs about deer, to the loud accompaniment of ABBA. Through the steamed-up windows we caught fleeting glimpses of darkened villages with a distinctly medieval feel to them; streets were eerily deserted, although there were lights on in most of the houses. Names like Kosieczyn, Wojciechowo and Powodowo appeared and disappeared in the gloom. We passed an empty and deserted Babimost airfield, a plinthed Mig jet guarding its entrance as a stark reminder of the Cold War era of the not-too-distant past. However, to this day, our most vivid memory of that journey remains our driver's eccentric habit of throwing his red, bi-fuelled Volkswagen Golf into neutral just before every bend, and coasting round it!

WELCOME TO WOLSZTYN

At Wolsztyn, we were met and welcomed at the Wolsztyn Experience Society's House by our genial host, Howard. The house was a former bar, bought by the Wolsztyn Experience to accommodate participants in the footplate experiences. Its foyer was covered in photographs of Polish steam locos, several carriage destination boards and various numberplates and plaques bearing the Polish Railways PKP *('Polskie Koleje Państwowe')* eagle insignia. On the counter stood a fish tank and a locomotive air brake handle; on a shelf behind it was proudly displayed a large, realistic, lacquered cardboard model of a German 'Kriegslok' 2-10-0. In one corner stood a TV monitor with a DVD player and a selection of railway videos, books and magazines from all around the world.

After a brief run through of formalities, Howard informed us that he had spared us novices the horrors of working on the 05:17 to Poznań next morning – a temporary reprieve but one for which we were grateful. Instead, we were to work the 11:10 to Poznań and return in order to break us in gently. After guiding us to our rooms, 'showing us the ropes', and giving us instructions for the following day, he left us to our own devices. We retired to bed, full of a mixture of anticipation and trepidation.

Over time, we were to find that almost everyone connected with railways in Poland, if not quite everyone in the whole country, seemed to know, or know of, Howard; the mere mention of his name and production of his business card was like an Open Sesame. Or, potentially, a Get Out of Jail card which, thankfully, we never had to use. His achievements in preserving working steam in Poland are legendary throughout the world of railways, and in the town of Wolsztyn itself. (We were able to obtain a slightly more advantageous exchange rate in a local *Bureau de Change* simply by mentioning his name!) An MBE was awarded in 2006 for his unparalleled achievements.

Howard had an endless fund of stories, both from his time in the travel industry in the UK and from his exploits at Wolsztyn. He had a wonderful sense of humour and the skill of being able to communicate effectively at all levels, to use the language of the business world. A straight-talking, no fuss, no nonsense bloke, he had the uncanny ability to think on his feet and get things done. No reasonable request to him for anything out of the ordinary would go unfulfilled. With repeated visits to Wolsztyn, Howard got to know and trust us; he knew that we would be flexible and change plans if asked. On several occasions, he had to go away on business and left us in charge of the Society House. Which leads me to …

THERE'S A CAR IN THE HOUSE

Howard was almost lost for words one night in 2005, when he left Julian and me in charge of the house. We had enjoyed a convivial beer or two at the table in the front garden but moved inside as it began to rain. We had been inside for no more than five minutes, when we heard an almighty crash of splintering metal. Looking outside, we saw that a silver Fiat Cinquecento had embedded itself in one of the pillars holding up the balcony! At best, it would have missed us by inches. At worst …

After making sure that the driver of the car – a young

girl – was unhurt, and making her a cup of tea, (well, why not?) we realised that we would have to contact Howard. But there was a problem – we had no phone number for him. Fortunately, we did have the number of his brother Trevor, who handled the UK side of business. This was in the relatively early days of mobile phones but, luckily, Julian had a work one which would make international calls – mine would not. Eventually, Trevor answered and gave us the number we needed. The ensuing conversation went something like this (expletives deleted):

'Good evening, Howard. We think you need to come back here because there's a car in the house.'

'Very funny, have you two been drinking?'

'Just a beer or two. Honest, a car has just driven into the pillar out the front.'

'What? Very funny. Are you joking?'

'Er, no.'

'How did that happen?'

'We've no idea but you really need to come over, our Polish is crap and the Police have just arrived.'

'OK, I'm on my way but you had better not be pulling my leg!'

Some ten minutes later, Howard arrived, driven by his partner Alex. (Ironically, Howard could drive a steam locomotive but not a car!) After talking to the Police, he was able to find out that the driver of the car claimed to have swerved to avoid a bicycle and, with the road being wet, had lost control of her little Fiat. Half an hour later, a tow truck arrived to take the car away and that was the last we heard of the matter. This near-disaster was just one of so many memorable things which happened to us in Poland. The majority were railway orientated but this particular non-railway one certainly had a lasting impact!

Pm36-2 waiting on the Exit/Entry road

The Author, firing (J.E. Maddocks-Born)

Wearing his 'obligatory' PKP cap, Julian performs the Ceremony of the Cleaning of the Wheels and Valve Gear

A smiling Janus returns from a trip along the footplate

2

A BAPTISM OF FIRE (AND WATER ...)

LEARNING THE ROPES

The next morning, following our adventures on the Berlin-Warsaw Express, and after a cooked 'English' breakfast with a Polish twist (salad!), Howard, Julian and I set off on the first proper step of our footplate experience. Howard led us to the engine shed, or *'Parowozownia',* where, besides the steam locomotives, we noted the fascinating sight of an ancient East German Trabant in the car park.

The depot was a place of wonder which brought so many memories flooding back of our local Wolverhampton boyhood steam sheds: Oxley, Stafford Road and Bushbury. During the 1960s, we had both spent many happy hours just soaking in the atmosphere of these working steam depots, complete with all the grime and grit that went with the territory 'back in the day'. Not to mention the constant fear of summary ejection! For now, there was no time to linger.

Standing on the shed's exit road, we found our loco for the day, waiting to go to the station to back onto its train. Pm2-36 was a shiny bright-green Pacific, sporting the name '*Beautiful Helena*' on one smoke deflector and '*Piekna Helena*' on the other. Naming locomotives is an alien concept to the

Poles, indeed to most other countries, and this one had only been named since being taken over by Wolsztyn. One of a class of two, it was built in 1936 for Polish State Railways. Examining her in detail, I thought I could detect a hint of French design in her appearance. Gazing up at this impressive machine with its twin-bogied tender, it struck us both as being considerably larger than anything ever built for Britain's railways. The Pacific had already done a 101-mile round trip to Poznań on the 05:17, returning on the 08:55. Howard chatted with the crew in fluent Polish, whilst Julian and I set off on foot along the road to Wolsztyn station, a journey of about half a mile.

The station was a grim Soviet-era building which has only very recently been modernised. The entrance hall felt more like a church, the ticket office like a confessional with the woman behind the counter half-hidden behind a net curtain. We felt a little conspicuous in our clean clothes, notably the check shirts which we had been forewarned were virtually obligatory. We carried virgin safety gloves, wore shoes with built-in toe protection and felt ready(ish) to work the 11:10 cross-country train to Poznań, some 50 miles and two hours away. This was the last time we ever walked along the road; from then on, just like normal Poles, we walked alongside the running lines if we ever had to get from the depot to the station, or vice-versa!

On the way to the station, our loco passed us, reaching the station before we did. By the time we arrived, it stood simmering at the head of a rake of four drab olive-green articulated double-deck coaches reminiscent of the Communist era. We climbed into the biggest steam locomotive cab

I had ever set foot in – large enough in which to hold a metaphorical dance. Howard introduced us to our moustachioed, check-shirted driver, Zenon, and our mustachioed, check-shirted fireman, Marek. Both wore scruffy old jumpers over their shirts; Marek sported a paunch, but Zenon was built like a rake handle. (Zenon died in 2010, Marek more recently.) We were impressed by the fact that the loco sported electric lights – both headlights and tail lights – and even had lighting in the cab, provided by 230-volt bulbs. Especially impressive were the lamp shades made from old beer cans! The first thing we did was to rub our gloved hands in the coal on the tender shovelling plate in order to make them look a little less obviously new.

Marek immediately sold each of us a PKP baseball-style cap, which he insisted was compulsory in the cab. (I still have that cap.) 'Good price, I give you good price' seemed to be his permanent mantra, sounding more as if he were in an Eastern bazaar than on a European steam locomotive. Only later were we to discover that Marek would have sold us the loco if we had enough złotys on us! We quickly nicknamed him 'Marek the Shop' as he constantly, but good-naturedly, badgered us to buy a whole raft of 'souvenirs' which he had stashed in various lockers all over the loco and tender. He even offered to sell us the small piece of paper with our schedule and passing times on it, which was held in a small clip frame close to the regulator. Zenon said little but shook his head, smiling ruefully from time to time at Marek's antics. (The following year, Marek sold me a cast aluminium PKP plate which he swore was ex-loco and genuine. Several years later I discovered that some of the crews were casting replica plates for mugs like me in home workshops!)

HOWARD ACCOMPANIED us on the footplate, driving the train away from Wolsztyn and making it look simple and effortless. As we got under way, he gave us a few hints on being in the cab but the experience was so overwhelming that he may as well have saved his breath. After passing a Prussian style signal box on the left, our line veered off to the right and began to climb sharply, curving left in a semi-circle to cross over the line to Leszno and towards our first stop at Tłoki. Howard gave us a running commentary on what he was doing, most of which went in one nervous, over-excited ear and out of the other. Our 'training' consisted of a few short phrases:

> *'That's the fire, it's hot.'* (Our sole Health and Safety briefing in ten years of footplating ...)
>
> *'That's the regulator, it makes us go.'*
>
> *'That's the reverser, a bit like the gearbox on a car.'*
>
> *'That's the brake, it makes us stop.'*
>
> And, finally ...
>
> *'That's the injector, it puts water in the boiler. Have a good day, lads!'*

And so saying, Howard bailed out and into the shedmaster's waiting car, leaving us rookies to the tender mercies of the Polish crew. Here ended our training! Neither of us had had much real footplate experience beyond the odd illicit

British Railways or National Coal Board cab ride in the days of steam. We thought we'd break ourselves in gently and simply observe the routine in the enormous cab of our Pacific. Julian stationed himself behind the driver's seat, which was on the right, and I stood behind the fireman's seat, on the left. It came as a major shock, therefore, to Julian when, as we pulled out of Tłoki, our train began to accelerate away and Zenon eased himself out of his seat, motioning Julian towards it, saying tersely:

'You drive.'

'But I don't ...'

'Is no problem. I tell you.'

At the same time, Marek ushered me into his seat. I thought that I would watch what Marek did and continue to acclimatise myself whilst Julian sweated. I tried to take in the excitement of being in the cab of a hardworking steam loco. To sum up: it was very hot, very noisy, very uncomfortable, very draughty, very smoky, very dusty, very dirty, and very smelly. I can do no better than to quote an American course participant, Joe Monty, writing in the bi-monthly American *Live Steam Magazine:*

> Everything rattled, the whole machine vibrated, steam roared out of the cylinder cocks, the cab shook, and we were under way with the cacophony of banging metal surfaces, roaring fire and hissing steam that typically fills a locomotive cab. The heaving and grinding of

> metal parts being pushed and pulled by the pistons combines with the accompanying shaking of the floor going through your body, making for an intense and thoroughly addictive sensory experience.

For me? A boyhood dream come true.

LEARNING TO SWEAR. IN POLISH

Interrupting my thoughts, Marek pointed to the injector: '*Jeff, water on.*' Trying to appear as nonchalant and confident as I could, I turned the handle smartly anti-clockwise, as I had seen Marek do. Or so I thought … All hell broke loose; every pipe in the cab rattled, banged and vibrated as if a host of devils was inside them. Steam issued from a multitude of places, hot water spat everywhere, bubbles of water appeared where none should be, and Marek shouted something incomprehensibly Polish at me.

In one split second, I learned two very important lessons that would stand me in good stead for a decade: firstly, that's *not* how you turn on the injector, and, secondly, '*curva*' is an extremely rude Polish word ... We heard it a lot over our ten years on the footplate! With a lot of sign language and exaggerated gestures, none of which, in my embarrassment, I really understood, Marek demonstrated how to put the injector on. Properly.

There was a brief interval to relax and enjoy the scenery, take a few photographs, admire the deer, the nesting storks, the huge eagles that flew regularly across our path. Meanwhile, Julian struggled with his first lesson in the intricacies of shutting the regulator, winding out the reverser to 60% and allowing the train to coast for several hundred yards, before braking gently to a halt in more or less the right place, marked by a white cross, at a station called Rostarzewo.

About 200 yards before the station, Marek's terse '*Jeff, coal to fire,*' accompanied by a sweeping gesture towards the shovelling plate, brought me back to reality. Frantically, I shovelled coal from the tender across the yawning gap to the firebox door, Marek showing me with expansive gestures how to spread it round the sides of the firebox, in the back corners and under the door. I managed to throw a few random rounds into the dancing flames of the scorching-hot inferno but as our train lurched to a halt, I caught the shovel on the backhead and sprayed its entire contents over the footplate, missing the firebox completely. Laughter all round. I joined in – through gritted teeth.

Whilst we were stopped, Zenon leaned out of the cab door to take a few pulls on his latest cigarette – it was lung cancer that took him some years later. '*Odjacz,*' he shouted, accompanied by gestures which left us in no doubt that it meant '*Right Away.*' Julian pushed hard at the regulator – nothing happened. Bracing himself against the cabside, he pushed much harder and it opened with a bit (OK, a lot) of a struggle. The Pacific promptly lost its feet, slipping like mad, a tall roaring column of black smoke issuing from the chimney. Zenon dived to close the regulator. '*Piwo!*', the crew shouted in unison. Probably my third word of Polish. Marek explained: '*Wan sleep eez wan beer.*'

In a second, the grin was wiped from my face as Marek raised his eyebrows, nodding wordlessly towards the injector.

I opened it a little more cautiously … same result: steam and hot water spitting everywhere, lots more *curvas*. Once again, Marek patiently, and with more gestures and Polish words, showed me how to put the injector on properly, just cracking it open then waiting for the nut to stop bobbing up and down, before turning the handle all the way round to the six o'clock position. Simples! If only.

Quickly and efficiently, the Pacific got back into its stride. Mopping my brow and getting my breath back from the sudden physical exertion, I again made the fatal mistake of relaxing in my seat, trying to observe the agricultural landscape outside the cab. I had just spotted another large eagle swooping across in front of our train and was reaching for my camera to photograph a man ploughing with a horse, when: '*Jeff, water on*', commanded a grinning Marek. Same result again ...

Not quite so patiently, he went through the pantomime of how to do it and once again, between us, we got water into the boiler. Turning the injector off was a lot easier but was rapidly followed by the 'coal-to-firebox' gesture. This time I managed to get all the coal into the box, bracing myself and waiting for the lurch as the train came to a complete halt at Rakoniewice. Progress? Well, I thought so until Marek grabbed a long implement from the tender and reorganised my fire, accompanied by a theatrical tutting and shaking of the head, which I think were in jest.

A SIMILAR PROCESS was repeated as we stopped at station after station, with names like Drzymałowo, Ruchocice and Granowo Novytomiskie. We paused for a little longer at Grodzisk, admiring the way that the place name was spelt out on the signal box in the manner of a Scrabble word. Here, there were signals so the crew took their rightful places in their own seats whilst we sat on little dickie seats which unfolded from the cab doors. It was explained to us, in a mixture of sign language and pidgin English, that this had to be done wherever there were trackside signals because we were not allowed to drive in such places. As we would discover, the implementation of this rule depended on the level of confidence the crew had in you.

Grodzisk has two signal boxes, the second of which had a rather attractive signal lady in attendance, standing at the window of her signalbox with a flag extended to denote that the adjacent level crossing was clear for our passage. Marek directed several obviously lewd comments and gestures at this woman as we passed – some languages are universal. After leaving Grodzisk, Marek peered into the firebox, pretended to glower at me, tutted a bit and effortlessly threw in a few rounds of coal, once more demonstrating where it should land. I even got the injector to pick up on my own. Once. At every station, great clouds of oily black smoke rolled back over the train and nearby countryside. I knew this could not be the most efficient way of firing, and certainly not the most environmentally friendly, but lacked the knowledge – and the Polish – to say or do anything about it.

Three quarters of an hour later, and after several more stops, we crossed over the electrified Berlin-Warsaw main line just after the village of Wiry, a few miles from our destination of Poznań. Our train rolled gently down the bank to join this line near Luboń. It seemed that our tasks were over for the time being, as the crew reclaimed their seats and we became mere spectators, and very relieved ones at that. Marek nonchalantly put more water in the boiler but signed that there was no need for any more coal. Some two hours after departure, and after stopping at two stations in the suburbs of Poznań, we rolled into Platform, or '*Peron*', 4b of Poznań Główny station where, in time-honoured fashion, I leaned out of the cab window to watch our passengers disembark and hurry on their way towards Poland's fifth-largest city. Not one of them spared our locomotive or us a second glance – certainly none of them said '*dziękuję*' ('thank you') as I used to do to the drivers of the Castles and Kings which once hauled me from Wolverhampton to Paddington.

DRINKS IN POZNAŃ

Not long after we came to rest in the station, Marek took me by the arm, grinned at both of us and said, '*Jeff, Julian, please, you go shop, buy cold drink, maybe Coca-Cola, maybe Sprite, for Marek and Zenon*'. We scurried off to find a shop selling drinks, whilst he climbed down from the cab to uncouple our machine from its coaches. With the station being such a large one, it didn't prove too difficult to find the requested cold drinks. We treated ourselves to the same – drinks which hardly touched the sides as they went down! It quickly became customary to buy cans, Mars bars or even cigarettes for the crews at major stations without being asked. On very hot days, ice creams went down a treat too; it all helped to keep the crews onside.

Having completed our first footplate ride, we were, inevitably, somewhat grubby in hands and face. Wearing check shirts and PKP caps, we must have looked like genuine railwaymen as an old lady stopped us and asked us a question, probably '*What time is the next train to Warsaw?*' '*Sorry, English*', we had to say. This subsequently happened to us on several occasions so I learned the words '*maszynista angielski*' ('English engine driver') to excuse ourselves. We got some strange looks over the years ...

Poznań station was a hive of activity. We were aghast at how many of the passengers eschewed the use of the subways between the platforms and instead walked casually across the tracks to the opposite one. This would be done right under the noses of the station staff. As in much of Europe, Polish platforms are lower than ours – but no station staff, not even the baton-wielding Railway Police ('*Straży Ochrony Kolei*') batted an eyelid at this flagrant disregard of the many warning notices. Visible just outside the station was a wide road crossing with barriers for road traffic; they may have stopped cars but they did not deter foot passengers from dicing with death!

The station was divided into roughly two halves: the long-distance side and the local services side. (In Poland, fares are lower on slower, stopping trains than they are on express services run by PKP's *INTERCITY* division) Each express train that came in was announced over the station Tannoy, the announcer reeling off the list of destinations for the train in question and then reciting the coach numbers, together with the classification of each one. In this way, I soon learned some Polish numbers! Long distance expresses came and went, most having boards on the side declaring where the train had originated and whither it was bound; some had started from, or were bound for, other countries.

An especially exotic and evocative one proclaimed on the side:

Berlin – Poznań – Warszawa – Brest – Minsk –
Vitebsk – Sankt Petersburg

Some expresses divided or had coaches attached by one of the station pilots, something no longer seen at home. Also not seen or heard at home was the once-familiar sight and sound of a wheeltapper striking his long hammer on the wheels of waiting carriages. We had our first glimpse of the Polish electric locos which bore a strong resemblance to the old British Class 83s, as well they might, since the first ten were built by English Electric at Newton-le-Willows. If you were in the right place, it was possible to see examples of working Polish steam, diesel and electric locos side-by-side.

BRIBERY AND CORRUPTION PART ONE

A few minutes after buying the cans of drink and handing them over to the crew, Marek took the controls, drove us forward, and then reversed over a network of points and crossings to the nearby motive power depot to take water, clean the fire and oil round our Pacific. Pm36-2 was too long to be turned on the depot turntable, meaning our return journey to Wolsztyn would be tender-first.

The whole shed area had a rundown feel to it, reminiscent of the end of steam in the UK. Outside the main building were several lines of withdrawn ancient, Eastern European-looking diesel locos, their matt two-tone green paintwork faded and flaking with rust. Nobody took any notice as we sneaked round the large maintenance area, which was surprisingly empty. It was just like 'shedbashing' in the old days back home, though we needn't really have sneaked round since we looked like proper railwaymen, complete with our PKP caps, courtesy of Marek. How he must have rubbed his hands with glee upon first encountering us two 'virgin' crewmen!

We were booked to lay over here for nearly four hours and then work the 17:00 return to Wolsztyn. Mr Fixit had other ideas. Disappearing for a quarter of an hour, Marek returned with a cheesy grin on his face and the news that he had managed to 'negotiate with' (i.e. bribe) the driver of the diesel on the 15:25 to Wolsztyn that we could swap turns and return at the head of the earlier train. He explained that this was because he had a young and attractive wife and wanted to get home early for some *'pumpa pumpa'*. In case we could not work out what this was he gave us a graphic demonstration, over which a veil should probably be drawn. But … *it would cost* us *20 złotys apiece*, just over £3 in those days, probably nearer to a fiver now. It seemed like a good plan as we had no idea how we would pass such a long time. Naively, we coughed up and, somewhat earlier than planned, our merry band steamed back to the station to couple up to the stock of the 15:25.

The orange-clad station shunter was not at all amused by our unexpected arrival because our train was now made up of four clean, smart Cross Country single-decker coaches, rather than the grim double-deckers that we had hauled to Poznań and which we later came to christen 'cattle class'. An argument ensued with many heated words, plenty of gesticulation and lots of *curvas,* all of which Marek nonchalantly brushed aside, as well he might, being 40 złotys richer! We doubted whether Zenon would get a 50% cut of the proceeds.

MY FIRST DRIVE

Before departing Poznań, and at Zenon's behest, Julian put plenty of coal on the fire ready for the journey ahead; thick black smoke billowed all around the station, drifting towards the nearby city streets. It was now my turn to drive so I took my place on the dickie seat behind the driver. Exactly on time, our guard blew her whistle, Zenon shouted '*Odjazd*' across the cab to Marek, and we left in a cloud of steam from the cylinder drain cocks and trailing thick grey smoke behind us. Once we had left Poznań behind, we made our way along the main line, stopping at the suburban stations of Dębiec and Luboń, and thence onto our cross-country route, coming to a halt on the steep, curving gradient at Wiry. Here Marek, who, as is the Polish custom, was now the official driver of our returning train, ushered me into the driving seat.

My sole experience of driving a steam locomotive was an unofficial cab ride on the National Coal Board in the late 1970s: a simple three-position lever reverser – full forward gear and only full regulator ever being used. In layman's terms – flat out! Now, with no real instruction other than observing Julian in between bouts of shovelling and ongoing battles with the injector, I was in charge of a powerful Pacific, considerably larger than a Duchess, and tender-first. Whilst I ought perhaps to have been excited, I actually felt a mixture of disbelief and terror at the responsibility!

Driving backwards meant that I had to kneel on the seat, craning my neck awkwardly out of the cab window, but having to climb off the seat and stand up every time I had to wind the reverser back, open or close the regulator, or sound the whistle. Each time I stood up, I was guaranteed to bang my head on the roof-mounted battery box which controlled power for the loco and cab lights (a good job I had bought a PKP cap!). Operating the whistle had to be performed literally dozens of times as the Poznań-Wolsztyn line is criss-crossed by a multitude of roads, farm tracks, dirt tracks and foot crossings, all requiring a statutory toot. Failure to do so in time was rewarded with simultaneous cries of '*Vissel!*' from our crew members.

Like many railway enthusiasts, I had read articles by knowledgeable railway writers, which described cut-offs being gradually notched back in very fine increments to as little as 15% or less. No such finesse here – we started in 60%, pulled back roughly every 100 yards to 50%, then 40%, then 30%, where cut-off remained until the regulator was closed. In the fullness of time, I learned to notch up simply by listening to the beat from the chimney, rather than waiting to be told, but for this first drive I just obeyed instructions.

The regulator itself had a handle which resembled a bird's beak and, as Julian had found, was very stiff, requiring me to brace myself against the cab side and push for all I was worth. The regulator plate had a series of teeth, the regulator had a lever with a handle; this handle had to be pulled up then dropped into the appropriate place on the teeth. I've yet to see a Polish loco worked on full regulator, normally just over half suffices. (I tried on several occasions to sneak it over a bit further but always got caught out!) It was explained to me that I had to watch the steam chest pressure gauge and not allow it to rise above one megapascal – *"Wan maximum,"* as Marek put it. To this day I'm still not sure what a megapascal is in terms of pounds per square inch or even atmospheres – and I really ought to know. I gradually realised that, although it took a lot of strength, it took relatively little skill to drive the big Pacific provided that I didn't open the regulator too wide. If I did, the loco slipped, I was reminded of the '*Wan maximum*' rule and fell deeper into beer debt. On the other hand, if I failed to hurry things along, the crew shouted the Polish equivalent of 'Get a *curva* move on' at me. As a mechanical exercise, the whole performance was simple; physically and mentally it was exhausting, pressurised and nerve-wracking.

Making our loco go was the relatively easy part; making it stop in the right place was not. For me, the hard part was understanding the 'feel' of the air brake and stopping in the right place. Braking was only achieved with massive help from Marek; I just could not get the hang of 'putting some brake in' then backing off on the handle and being confident that braking was still taking place. As a car driver, this was counter intuitive. It seemed to take ages for the brakes to 'bite' then, all of a sudden, we would be stopping short and the air would be hastily dumped by Marek, along with a *curva* scolding. If we did stop short of the white cross then not only would Marek pass rude and *curva* laden comments, but passengers would glower at me and the guard – a short, dumpy woman with bright red hair – would pass rude comments. At every stop, she kept up a running commentary with the crew, no doubt casting aspersions on my masculinity, knowing that I would not understand a word and probably not caring if I did!

My beginner's ineptitude with the brakes seemed to have an adverse effect on the air pump, situated alongside the smokebox. After one particularly amateurish effort at stopping at yet another wayside halt, the crew declared it as '*kaput*' so, without further ado, and as we began to accelerate away from the stop, Marek climbed out of the cab through a front door which I had not noticed before, shuffled along the running plate and began to attack the offending pump with a large spanner and a large hammer, leaving me to drive unaided. Julian called me over to witness this, in spite of the fact that it meant I had to desert the driver's position. I simply could not believe my eyes but somehow managed to maintain enough composure to photograph this incredible event which took place at a speed, I would estimate, of around 30 mph. Even though I was supposed to be driving, Zenon appeared unconcerned by the sight of me abandoning my post to take the photograph.

On returning to the cab, Marek seemed confident enough that he had fixed it. *'All in a day's work,'* his expression seemed to say. However, after the next stop, at Ruchocice, Zenon expressed his dissatisfaction with Marek's efforts and proceeded to perform exactly the same feat! Another photograph was taken as I was convinced that no one would believe our story. In this fashion, we eventually pulled into Wolsztyn, with Marek at the controls for the final mile or so, spot on time in spite of my best (worst?) efforts. We stayed on board whilst uncoupling, shunting our stock and returning to the shed for disposal, coal and water. Not knowing quite what to do and feeling slightly redundant, we hung around watching the goings-on ... but not for long as we were quickly initiated into the longstanding Wolsztyn custom of cleaning the white-walled wheels and the red painted motion, outlined in white. This was an unpleasant task in such close proximity to the hot, hissing and steamy locomotive, but a way of showing willing. To say that our first trip on the footplate had been eventful and a huge eye-opener would be a massive understatement. We slept well that night, safe in the knowledge that we were not working the 05:17 the next day.

Poznań Arrival

The Dream Team: 'Mad' Andrzej lights up another cigarette whilst Irek relaxes in the sun at Zbąszyńek

Ol49-69 leaves Wolsztyn with the 15.58 to Poznań

Konotop

3

FINDING OUR FEET

BUM AND SPIT

The following day was different in almost all respects from our 'baptism of fire and water'. We had a different crew – known to all as 'Bum and Spit' – and a different loco, this time a 2-6-2 tender loco, Ol49-111. In case you are wondering, 'Bum' got his nickname from his habit of leaning out of the cab window whilst kneeling on his seat, his ample posterior thrust into the cab. And 'Spit'? He too had a habit …

This time, we joined the crew at the depot and rode light engine to the station, where we backed onto yesterday's double-deckers to form the 11:10 to Poznań. Our new hosts were totally indifferent to our presence on the footplate, making all requests by means of gestures and scowls, but still allowing and expecting us to do all that was done the day before as if we were seasoned footplatemen. The difference from the previous day was that there was no attempt at explanation or teaching. 'Bum' was short and stout, sporting a Viva Zapata moustache like a Mexican bandit. We never once saw him smile. Janus, the senior driver at Wolsztyn, referred to him as 'Mr No Happy'. ('*You know why he no happy, Jeff?*' he asked me one day, '*He no have wife, he live with Mama.*') His real name was Ryszard and his crewmate was Stanisław, an impatient man. We would meet him again.

Although Ryszard let me fire, he was quite incensed by my efforts, attacking the fire frequently with various implements dragged out from beneath the tender, accompanied by a staccato litany of *curvas,* then rearranging it into an entirely different shape from what I had been shown by Marek the previous day. And as for my efforts with the injector, let's just say he was *curva curva* unimpressed, although personally I felt that I did much better than the day before.

With the Ol49 being a shorter loco than the previous day's Pacific, it was turned at Poznań so I did not have to drive back tender-first. The 2-6-2 seemed an easier machine to drive, less inclined to slip and with lighter (but certainly not light) controls. With the exception of a few minor slips, which I was quickly able to control without any prompting from the crew, the journey back to Wolsztyn was relatively uneventful and we began to relax a little, feeling that we had at least a rough idea of what we were doing. We came to the conclusion that the crew's apparent unfriendly indifference was not personal. Looking at things from their point of view, they had a train to run safely and to time and did not need the 'assistance' of cack-handed, rank amateur beginners. In spite of the lack of warmth from the crew, bordering at times on what felt like downright hostility, we still enjoyed ourselves and put the less-than-friendly atmosphere down to experience. On the other hand, the crews made a good living from the Experience and were rewarded with generous weekly tips, which Howard collected from course participants, so they had good reason to tolerate our presence.

OUR FIRST FREIGHT

Wednesday saw us again avoiding the 05:17 but still having to make an early start, so we were in for a long day. In the goods yard at 07:00, we joined Monday's crew of Marek and Zenon on the footplate of Ty45-379, a squat 2-10-0 sporting smoke deflectors and coupled to a 'bathtub' tender. After shunting the yard for a while, the wagons carrying traces of snow on them, we assembled a short freight of three bogie vans which we took as far as Wolsztyn station, to await the 'Right Away' from the signalman.

Once Marek stopped trying to sell us anything which was not bolted down, an orange-clad shunter joined us on the footplate. Introductions were made, hands were shaken all round and we set off for Tuchorza, a few miles up the line towards Zbąszyńek. As soon as we were out of the confines of the station, I drove the short distance to our destination. With such a lightweight load, little effort was required of our loco and Julian had to do very little firing. Annoyingly, he seemed to master the injector rather quickly!

As Howard had warned us, our 2-10-0 was fitted with Trofimov valves, which meant a different routine when closing the regulator from what we had experienced the previous two days. For those who are unaware, Trofimov is, and I quote, *'a springless, pressure-compensation piston valve system'* which is supposed to aid freer coasting. When the regulator was closed, the reverser had to be wound out to full forward gear. You had to listen for the valves falling onto their seats with an audible 'clunk', then wind the reverser back to mid-gear. Then, when the train had come to a halt, the reverser would be wound out to 60%. Zenon did his best to explain with sign language and faltering English but I failed to understand properly. At one point, the reversing wheel slipped out of my hand, spinning wildly round and catching my right arm with a blow so fierce I feared it was broken. Fortunately, I escaped with just a massive bruise on my forearm and severely wounded confidence! Ol49-7 was fitted with the same type of valves; the one time I drove it with Marek, he insisted on taking over each time the regulator was closed as he obviously did not trust me with the Trofimov routine. That time, only my pride was hurt; my arm was spared!

The crew did all the work as we shunted the vans into the station yard at Tuchorza. Delicate touches of the brakes and regulator were required and us raw trainees would definitely not have had the finesse for the task. From Tuchorza, a gentle, light engine, tender-first amble took us back downhill to Wolsztyn to await further orders. Dropping off our shunter in the station, we headed off light engine past Tuchorza once more, joining the electrified main line at Zbąszyń. I had to metaphorically pinch myself when I realised that I was actually driving, albeit a light engine, on the Berlin-Warsaw main line! A little further on, we came to Zbąszyńek where we were routed into a huge marshalling yard, easily as big as anything ever seen in the UK. Before reaching the yard, we passed a preserved 0-8-0 tender engine, parked in a short siding. We waited what felt like an eternity. Marek made a few half-hearted attempts at selling us 'souvenirs' but the crew were quite happy to pass the time chain-smoking, before hooking onto two empty, long-wheelbase, bogie coal wagons.

With Julian now driving, our tender-first return was enlivened by the sudden appearance on the line ahead of a herd of deer. Instead of heading away from our train, they insisted on running ahead of us between the rails. Braking hard, we narrowly missed a couple of the terrified animals as we rolled gently downhill towards Wolsztyn. We deposited our coal wagons in the goods yard, before heading across to the depot to dispose our loco. This was to be the only time we ever rode on Ty45-379. From the following year, it was dumped either in the goods yard with other 'wrecks' or left in the depot yard to rust.

OUR FIRST TRIP TO LESZNO. WITH THE MAD LOCOMEN

Back at the Society house, there was time for a scrub down before our first trip to a Polish supermarket – Netto – to purchase a few things for lunch. In itself, this was an interesting experience; there seemed far less choice than back home, and food and other goods were not particularly enticingly displayed. We did, of course, have the minor problem of speaking no Polish when the cashier addressed us, not to mention wrestling with the złoty. Inevitably, in our embarrassment, we handed over a large note and ended up with pocketsful of 'shrapnel'.

After a lunch of local bread and *sopocka* (cooked meat), we donned our uniform of checked shirt, faded denims and PKP cap, and walked out in the sunshine to the depot. No need to sneak around as in our teenage years, no fear of being thrown out. This was our first look inside the half-roundhouse; in a second, we were transported back to the 1960s and our 'shed bashing' days. All the atmosphere was there, all the tools and paraphernalia of the steam age. An Ol49 simmered quietly on shed, hissing and smoking away to itself. The depot was like a time machine, transporting visitors back to a different age, but a proper look would have to wait.

Walking like real railwaymen alongside the track to the station, we joined the crew of the 16:45 service to Leszno, some 28 miles away. The loco had already worked light engine as the 14:00 to Zbąszyńek, leaving there at 16:00. We were reunited with our big green Pacific and introduced for the first time to the craziest pair of footplate crew you could ever meet anywhere in the whole wide world – Andrzej and Irek, the latter a dead ringer for the comedian Alexei Sayle. Any notion we had of footplate work being serious was quickly dispelled as Andrzej ran through some of his vast repertoire of practical jokes. These ranged from placing a large piece of coal on the driver's seat just as he was about to sit down, to tying the shovel to the tender brake handle or closing the firebox door just as a shovelful of coal arced its way towards the fire. His timing was impeccable, born out of many years of practice on unsuspecting novices. We quickly nicknamed him 'Mad Andrzej' – a name which has stood the test of time.

In later years, as Andrzej got to know us better, he would extend his range of tricks to setting our clothes on fire with his cigarette lighter, or hitting you on the foot with a large hammer – after first checking that you were wearing protective shoes! On more than one occasion, he disappeared out of one side of the cab – whilst the train was in motion, of course – and reappeared on the opposite side to give the unsuspecting driver, who would be concentrating fiercely on the road ahead, the fright of his life. On one occasion, I recall him standing on the top of the tender bombarding us with pieces of coal – whilst we were on the move! One year, as we approached a red signal on the main line near Poznań, there was a freight train on our line, directly ahead of us, so Andrzej and Irek pretended to bail out of opposite sides of the cab, climbing halfway down the steps.

Meanwhile, Irek insisted that we all duck down at each bridge we passed under. The remarkable thing about this was that he knew to the split second exactly where the bridge was without looking out of the cab. There was a laugh a minute, all to the inevitable accompaniment of chain-smoking by both crew members. Although there was much laughter and merriment on the footplate, it was still clear that the enginemen were in full control of the job every second, demonstrating more than once that they were ready to override us in a heartbeat. I fired all the way, both ways; Julian drove both ways, neither crew member showing any inclination to fire as long as I was willing to do so. Nor did they criticise my fire, thank goodness, probably because the loco just steamed along so they did not even bother to look at it. As for the injector, just when I thought I had got the hang of it, I muffed it. Cue another patient lesson …

Just before we arrived at Leszno for our first time, we noticed a large 3D model of an Ol49 perched on top of a wall. On the opposite side, there was a long and extensive freight yard adjacent to a large and ancient-looking motive power depot full of locos. Just as at Poznań, there were lines of withdrawn diesel locos of all shapes and sizes, almost all wearing the faded two-tone green colours of PKP. Unlike Poznań, however, there were several rusting hulks of derelict steam locos. And, again unlike Poznań, there were some withdrawn electric locos.

From Leszno we were to return to Wolsztyn light engine, tender-first since, as at Poznań, the depot turntable was too short for our Pacific. On this journey, we were employed as an unofficial taxi service for all kinds and ages of people of both genders. Leaving Leszno with in excess of a dozen bodies in the cab, the little firing that needed to be done was rather awkward – the Polish for '*Excuse me, can I just throw a few rounds on the fire*' was not in my vocabulary. Gradually, the cab emptied as we stopped, or slowed down, at a few crossings and dropped people off, including at least one very lucky schoolboy!

'WODKA'

Arriving back in Wolsztyn station, Andrzej introduced us to another local custom. Putting an arm around my shoulders in a fatherly way, he grinned.

> *'Jeff, you very good fireman, ok?'*
>
> *'Dziękuję,'* [Thank you] I simpered, falling straight into his trap.
>
> *'You see Netto over there?'*
>
> *'Mm, yes.'* The trap closed in …
>
> *'Jeff, you go Netto, buy two bottle wodka for Andrzej and Irek.'*

Julian declined my request to accompany me, flattering

me by telling me that I was better at foreign languages than he. Thus, I found myself in Netto for the second time in the day, this time sporting filthy clothes, filthy face, filthy hands and filthy feet, buying two bottles of *'wodka'*. Nobody in the shop seemed to notice or to care; they must have seen it all before!

THE 05:17 AND MORE AIR PUMP PROBLEMS

The following day, the 05:17 to Poznań finally caught up with us. Five o'clock found us bleary-eyed and shivering at Wolsztyn station, joining another new crew on the Pacific, this time the wonderful Janus, and his laid-back partner Koscianski. Janus had taken the trouble to learn English to a decent standard and was an absolute born engineman – with an unusual hobby. Keeping coypu! I know that he was absolutely gutted when a medical problem forced him off the footplate. The senior driver at Wolsztyn, he was always completely relaxed in the cab, completely at home and in his element. If he was really content with life, he would hum a Polish tune and even occasionally burst into song.

Standing behind me one day whilst I drove, Janus cocked his ear to listen to the beat of our engine, beamed from ear to ear and simply said, *'Muzyka'.* No other words were necessary. This marvellous crew could not have been more welcoming or more helpful and we had a wonderful round trip, both crewmen being excellent, patient instructors. With the cab being so large, every shovelful of coal had to be carried at least one step forward from the tender shovelling plate to the firebox but Janus demonstrated his mastery of the art of firing by straddling the loco and tender, loading the shovel and neatly flicking it to put the coal exactly where he wanted it – without even moving his feet. I felt more than a little helpless watching his performance but never once did he make me feel inadequate or inferior. Not even when I fluffed the injector. Again!

Yet again, we had air pump problems, and yet again, one of the crew – in this case Janus – headed along the running plate to the offending pump, returning to the footplate via the cab-front door with a huge smile on his face. That morning's ride into the sunrise remains an indelible memory but every time I took out my pocket camera, it seemed that the crew wanted me to fire or work the injector, which by now I had almost mastered, or perform 'blow-down'. This last operation consisted of raising a bar on the cab floor through 90 degrees, listening to the roar of steam blowing sludge from the boiler onto the track and then pushing the bar down again with a foot. Once more, I found myself driving the Pacific back from Poznań tender-first, this time on the 08:55. No easy job but, nevertheless, highly enjoyable as I felt that I had taken the first step along the long road to mastering it. I needed a lot of help with braking but Janus was an excellent tutor.

After a rest of a couple of hours, a quick wash-and-brush-up and a bite to eat, at half past one we walked to the station where we climbed onto the footplate of Ol49-69 to renew our acquaintance with the crazy duo of Andrzej and Irek. It became our custom during the course of the week always to shake hands with the crew and remind them of our names. They seemed pleased that we remembered their names; how could we forget them? Our remit was to work light engine, tender-first, to Zbąszyńek, where we would take water, before working the 16:00 from there to Leszno via Wolsztyn, returning light engine, as the day before. I fired almost the whole trip from Zbąszyńek to Leszno then back to Wolsztyn, doubling, as usual, as a taxi service on the final leg of the trip. By the time we climbed down from the cab at the depot, we had spent over ten hours on the footplate and were rather tired, putting it mildly. We had covered no less than 185 miles in the cab, a high percentage of it either driving or firing, a personal record which will never be beaten.

LAST TRIPS

And so, to our last day of footplating on our first 'Experience'. For the second day running, we had an early departure on the 05:17 to Poznań. The now-familiar crew of Marek and Zenon were in charge, this time of Ol49-69. Marek was obviously feeling the after effects of a heavy night – too much *pumpa pumpa* perhaps (more likely too much *wodka)* and was content to let me do all of the firing to Poznań without even looking in the firebox. I enjoyed this and began to feel like an old hand, especially when the injector played ball with me almost the whole way. I felt quite flattered when we reached Poznań at 07:15 and, whilst leaning on the cab window sill watching the crowds of commuters streaming along Platform 4b, someone asked me a question in Polish, obviously mistaking me for a real railwayman for the second time this week.

After a short break at Poznań for turning and watering the loco, we set off back to Wolsztyn. I fired all the way back as Julian and I had decided that he was better at driving –

especially at stopping – than I, and I was better at firing than he. It was tiring but very satisfying to be trusted to do all of the firing and all of the injecting. In his more alert moments, Marek referred to me as 'Jeff non-stop fireman' as he tried to sell me the working timetable sheet nestling in its receptacle next to the regulator. '*Interesant?*' he enquired. '*Nie*'. ('No', I had soon learned that word.) Definitely not, even at ten złotys, a reduction from the twenty demanded on a previous trip! Back on shed, we cleaned the rods and motion whilst the crew did the hard work of cleaning the fire. Shaking grubby hands with our very hospitable crew, we took our leave and congratulated ourselves on surviving and, equally importantly, enjoying five days on the footplate.

Howard had other ideas. As a 'beginners' bonus', we were rostered to work once more on the 14:00 light engine to Zbąszynek and the 16:00 return to Leszno via Wolsztyn. This time we had a new crew of Piotr, of blond, curly hair and film star looks, and Josef, a member of the shed staff who occasionally worked on the footplate. Both crewmen were unhelpful and intolerant; Josef, the fireman, allowed me a couple of turns on the shovel before deciding that I was useless and firing himself. Frankly, even though my ego was bruised, I was too tired from 100 miles of firing earlier in the day to protest, even though he did let me work the injector. Until I fluffed it. *Curva curva.*

The highlight of our return trip was an emergency stop when a farmer in a field alongside the line dared to drive his ancient Ursus tractor too close to the track for our driver's liking. (No fences of course, this is Poland.) Piotr swiftly removed Julian from the driver's seat and took over the reins. The hapless tractor driver received the highest number of *curvas* we heard that week! Even though we could have stayed on the footplate all the way to Leszno and back, breaking our daily mileage record, we had not been made to feel especially welcome so bailed out at Wolsztyn. We were absolutely shattered; it had been an early start and we had spent over eight hours in the cab, travelling 131 miles during the course of the day. We still had enough energy left though to race up the station steps to photograph the train departing for Leszno.

GOING OUT WITH A BANG

Our last day in Wolsztyn saw us without a footplate turn, so we finally got to enjoy a lie-in before a last, lingering wander round the shed and the dump of derelict hulks in the old goods yard. Waiting for the 15:58 to Poznań to depart behind Ol49-69, sleet began to come down almost horizontally so we watched the spectacle from the shelter of the cab of one of the many dead locos on the 'scrapline'. And that, we thought, was that, the end of a truly unforgettable week. We would have a last meal at the Europa Hotel, a quiet night in and an early night, ready for our journey home via Zbąszynek, Berlin and thence to England.

No chance! Howard 'ordered' us and two of the week's other participants to catch the 19:31 diesel-hauled Poznań service. We were to alight at Drzymałowo, which is basically an unlit, ruined, wooden hut in the middle of a forest, its single platform covered in knee-high grass and ferns. Unsurprisingly, we were the only fools to leave the train at this God-forsaken outpost. We groped our way through the pitch-dark forest towards the lights of a restaurant known as 'The Hut in the Forest'. Naturally, the menu was all in Polish and, equally naturally, no one spoke a word of English – but they did do a good line in miming the main ingredient of each dish! To make matters even more interesting, a 'band' consisting of an organist and a drummer entertained the locals with tunes like 'Tie a Yellow Ribbon Round the Old Oak Tree'. Several elderly couples danced lethargically to these classic tunes. We decided to forego the pleasure.

Just to add a little spice to matters, we had under an hour to order food of unknown ingredients, eat it, pay the bill, make our way back through the deep, dark wood, and catch the last train back to Wolsztyn. To this day, I have no idea what we ate, although it was almost certainly pork. Suffice it to say that Polish cuisine seemed to be basically a case of '*And what would Sir like with his gherkin?*' I exaggerate ...

We made our return train with minutes to spare, arriving at the station in time for Ralph from Birmingham – one of the initial supporters of the Wolsztyn Experience – to let off the loudest firework I have ever heard at the back of the station. (Howard seemed to hold a grudge against the restaurant and his instructions were that it was to be let off in the toilets but we chickened out of that one!) We watched spellbound as Ol49-69 steamed atmospherically towards us, its headlight cleaving a passage through the forest. Marek leaned out of the cab, saw us, pointed at me and shouted '*Jeff, non-stop fireman*'. I hoped for an invitation for a last ride but two other course participants were in the cab with the Polish crew, so we had to content ourselves with a ride in one of the dingy double-deckers.

BACK AT WOLSZTYN station, we hitched a lift in the cab to the depot. Whilst the loco was being bedded down for the night, I sat alone in the cab of the gently hissing engine for a good half hour, just soaking up the atmosphere of a live steam locomotive, the warmth and the smell. The distinctive aroma of a working steam locomotive would insinuate its way into any clothing which was worn in the cab. One particularly cold year, I wore an old fleece on the footplate and, rather than abandoning it in Poland, took it home with me. I hung it up in the cupboard in my flat and would daily relish the memories ingrained in its fabric. (I have long maintained that the essence of this aroma should be bottled and sold as perfume.) Carol, my then wife-to-be, once caught me with my nose buried deep in the garment with a big grin on my face, inhaling the memories. I will leave her comments to the reader's imagination.

Our first Wolsztyn Experience was over. And what an experience! We would be back – in my case eleven more times – though not always on the footplate. Altogether, Julian and I had travelled over six hundred miles in the cabs of four different steam locos. In the course of ten years on the footplate, I was to drive or fire almost five thousand miles in the cabs of mainline steam, electric and diesel, narrow-gauge steam and a narrow-gauge diesel railcar. Over the years, my Polish would improve to the point where I can now be confident that I have ordered two hamburgers, and not one double one, as happened the first time I tried. However, on my last trip, for some inexplicable reason, our party ended up with a pizza each as well as their burgers …

Pt47-112 storms past a lineside cottage near Krzycko Wielkie

Photographed from the cab of Pt47-112,
timber is loaded at Konotop

4

MORE FREIGHT

IN 2003, Julian and I flew to Berlin, this time separately, and again caught the Berlin-Warsaw Express to Zbąszynek. The only adventure, if it can be called that, was caused by KLM at Amsterdam, where I changed flights at the vast Schiphol Airport. The time between flights was just enough for me to make it to the Berlin plane. But not my luggage. I had to wait several days before that caught up with me. Sitting in the cab of Ol49-69 in Poznań station, the baggage company called from the airport. I tried to persuade them to bring the bag to the station but they could not be convinced that they would be able to find me. Telling them that we were in the cab of the only working steam locomotive in Poland, if not in Europe or in the Whole Wide World, failed to persuade them. My suitcase had been flown to Poznań Airport and was then treated to a 50-mile taxi ride to the Society House in Wolsztyn!

On the Wednesday of our week, we were not rostered to work on the footplate until 14:00, but, in the morning, I decided to wander down to the depot for a mooch around. Much to my surprise, Pt47-112 was engaged in shunting a rake of wagons around the goods sidings. Its train, which must have worked in from Zbąszynek, contained one loaded timber wagon, one empty timber wagon, one open bogie wagon and a whole rake of cement tanker wagons. The line to the Prefbeton works at Powodowo leaves the Zbąszynek line about a mile to the north of Wolsztyn. Empty trains have to change direction in Wolsztyn station, as do loaded ones, adding extra interest to operations. They were to become the last regular, timetabled steam-hauled freight workings in Europe.

The timber wagons and the bogie wagon were left in the goods yard sidings before the cement wagons were taken to one of the central lines – the middle road – the loco by now being tender first. I had time to take a few photos before the big Pt47 left the station for Powodowo, with empty wagons.

TIMBER

The following day, we were rostered to actually work on a freight train. At 06:45, we made our way to the goods sidings, joining Janus and Kochianski on the footplate of Pt47-112, the loco from yesterday. This massive 2-8-2 tender engine weighed over 100 tons just for the loco itself, the tender adding another 77. The loco and tender combined were a good ten feet longer than an LMS Duchess! It was also, of course, wider and taller than anything in the UK. Climbing up into the cab from ground level, it was difficult not to be in awe of its sheer size. This was our first hands-on experience of one of these impressive, 'modern' locos, this one being in a weather-worn grey livery.

Its spacious cab was fitted with what looked like a mechanical stoker, which I pointed at hopefully. Janus just smiled, gave the thumbs down and uttered that single, internationally understood word: '*Kaput*'. I picked up the shovel and opened the firebox door to look inside, realising what a lengthy firebox this powerful loco possessed.

We were coupled to three empty, open, bauxite-coloured bogie wagons. Our job was to take them along a virtually abandoned branch line to Konotop, a village to the south west of Wolsztyn, and return with loaded ones. At seven o'clock, we were signalled out of the yard and along the middle of the three tracks which run parallel to each other for a short distance in a southerly direction from Wolsztyn. The line to Poznań branches off to the right very soon after the depot, passing a derelict furniture factory and leaving two lines heading in the direction of Leszno. Within a few hundred yards, our line diverged from the Leszno line, passing beneath the Poznań line before entering another world, making our way gingerly along a rickety line where it was often difficult to distinguish track from undergrowth.

Our three empty wagons scarcely exceeded the weight of our tender, so very little effort was required from our locomotive as it trundled along at around ten miles an hour. We passed over several rusty and fragile-looking girder bridges (reminiscent of the Airfix Girder Bridge kits of our childhoods) and forced our way through overhanging trees and bushes. It was not long before, with a tinkling of glass, one of the cinderguards shattered; the one on the opposite side followed suit not long after. Janus drove, a broad smile on his face, humming and occasionally bursting into song; nobody could have looked more at ease with life. Kochianski, meanwhile, told me when to throw a few rounds in the firebox. Julian was travelling in the fireman's seat so he was entrusted with the injector. I was not sorry!

After about three quarters of an hour, we emerged from the jungle into the small town of Konotop. There had obviously once been a fairly extensive track layout here with several platforms; the whole place had the air of an abandoned wartime film set. There were still station name boards on the deserted, weed-strewn platforms, as well as a large water tower, though I doubted whether that would work. With the size of our tender, water was not going to be a problem!

Standing in a siding alongside the goods platform were four bogie wagons, identical to our three but loaded with timber. It felt like one of those shunting puzzles sometimes seen at model railway exhibitions. In order for us to deposit our three empties and hook onto the four loaded wagons, a whole set of complicated manoeuvres was needed. At one point our loco did an impression of a push-pull train, ending up in the middle of the seven wagons. Julian and I wandered across the tracks to watch and photograph proceedings.

After a leisurely hour or so, a few cigarettes smoked, a good natter and much laughter between the crew and the young man who had been responsible for loading the timber, we were almost ready to depart for Wolsztyn. This involved yet more toing-and-froing in the station yard, before we were ready to set off, tender-first, on our slow but uneventful return journey through the jungle, at the end of which we deposited the loaded wagons in the goods yard at Wolsztyn. The bucolic round trip had cost us two broken cinderguard glasses; no doubt they would be swiftly replaced by the shed staff.

THE RETURN OF THE BIRCH

There was time for a trip to Netto (something at which we were now old hands) to buy lunch. Then it was off to the depot to join up with the crew of Henryk – who was new to us – and Zenon, on what we were expecting to be a light engine working with Ol49-23 to Zbąszynek. However, we left the depot at 13:30 and headed straight for the goods yard, where we collected the four loaded timber wagons from Konotop.

Once we had left the confines of Wolsztyn station and headed tender-first past the remains of Skansen station, Julian was allowed to drive, under the supervision of Zenon, whilst I fired under the watchful (and critical) eye of Henryk. The second time that I was ordered to put on the injector, I reverted to my old habit of fluffing it. That's when I discovered that Henryk was much given to flapping and panicking. In addition to the traditional string of *curvas* to express his displeasure, he had a novel and rather idiosyncratic 'punishment' for offending footplate novices; if one of them were to do something which he considered wrong, he would rap them over their gloved knuckles with the besom which was kept for sweeping the footplate – and not gently! (For anyone not acquainted with the besom, it is the broomstick made of birch bark twigs that witches use for flying through the air.) Every time I fluffed the injector, Henryk would clobber me on the hand.

The four loaded timber wagons were deposited in the goods yard at Zbąszynek. After a wait of an hour or so, we left to continue the short distance to the station, from where we were to work the 16:05 to Leszno via Wolsztyn. Before hooking on to our short train, we took water beneath the electric wires in the station. The day had been a long, hot, sweaty one so, although we were rostered to footplate it through to Leszno on the two-coach train, and then light engine back to Wolsztyn, we elected to bail out at the latter, partly because Julian was to return to the UK the following day for business, but mainly because we were hot, sweaty, tired, hungry and thirsty!

On the Friday, I worked on the 04:16 to Poznań with an affable Londoner called Glen, who I distinctly remember refused to wear gloves whilst firing or driving, much to the amazement, incredulity and horror of our crew of Stanisław and Irek. Glen drove all the way to Poznań whilst I fired; we exchanged roles on the return journey with the 08:05 back to base. Whilst on shed at Poznań, Stanisław tried to sell me a red and white PKP plate such as those seen on loco cabs and tenders. I told him that I had already bought one from Marek, who had caught me at the depot the day before, making me an offer I could not refuse. Stanisław became quite cross and for the rest of the day addressed me with contempt as *'Jeff Bandit'!*

As soon as Glen had uncoupled us from our coaches, having been directly ordered to wear gloves for the task, we set off along the Zbąszynek line, leaving it just after Skansen and heading uphill to the Prefbet cement works at Powodowo, where we coupled onto a train of loaded bogie tankers. Glen and I left the footplate to take a few photos of the operation but Stanisław seemed in a hurry to depart, so we quickly returned to the cab, before we descended the bank to Wolsztyn where we parked the rake of tankers in the middle road. We were told that the loaded train would depart for Zbąszynek at 14:00 so Glen and I returned to our respective lodgings for a shower, a rest and lunch.

I left the Society House at around one o'clock and made my way out of town about half a mile along the Zbąszynek line, where I basked in the sunshine at the lineside to await the loaded train. Ol49-23 left the station punctually and attacked the bank, working harder than I had heard a Polish loco work before. As expected, the engine was working tender-first so that it would be facing the right way at Zbąszynek to work the 16:05 to Leszno.

The next year, we saw no freight and were informed that, as far as steam haulage was concerned, it had finished, a great shame because working on goods trains gave us a little welcome variety. And that, we thought, was that. However, 2005 had a very pleasant surprise in store for us, a major highlight of all of our Experiences …

THE FAKE FREIGHT

We arrived in Wolsztyn at the end of May 2005 to find that there was steam to both Poznań and Leszno, the latter involving a very long layover. After two days of the Leszno run, on one of which we took an express to Wrocław during our 'break', Howard came round in the evening and asked a favour of us. He had two members of the British Railway Press staying – we had seen them around but they had kept themselves very much to themselves. They had hired an empty freight train for the week to run on the Leszno line during a long gap in the timetable. After two days they were bored. Would we mind letting them have our early morning Poznań turn whilst we took the freight? Naturally, we jumped at the chance!

The following day, we caught the 08:51 to Leszno, where Andrzej was waiting for us on the platform. We walked together about a mile alongside the main line to the far end of the goods yard, Andrzej explaining to us in his half-decent English what the run would entail. The whole yard had a neglected air about it; knee-high grass and weeds hid the track. At the far end, we found Pt47-112 attached tender-first to a rake of eight bauxite-coloured, long-wheelbase, four-wheeled vans. We had quite a long wait in the cab but time would never hang heavy with Andrzej, who, whilst chain-smoking, kept us amused with his usual tricks and Marek tried to sell us anything and everything. Nothing new there!

The reason we had to wait so long became clear when a diesel loco passed us on its way to Wolsztyn, towing Ol49-7, which had been undergoing repairs at the nearby depot. The first section is a long one and the convoy was travelling slowly, making our wait a lengthy one. When the colour-light signal cleared, we set off at a brisk pace as far as Wloszakowice, where we stopped for our 2-8-2 to run round its train. Once we were coupled up and the brakes tested, we set off back towards Leszno, our brief being that whenever we saw what looked a likely spot for a runpast, we were to tell the crew, who would drop us off at the trackside, reverse up the line a hundred yards or so, put on Andrzej's celebrated 'Big Smoke' and thunder past us. This they would do at each spot as many times as we asked!

Our first request stop was part-way up a bank, next to a yard full of horses. We alighted and the Pt47 duly reversed down the bank a hundred yards or so. Once we had waved to signify that we were ready, it stormed noisily past us, smoke and steam wreathing the smokebox. The horses in the stable yard stared nervously as a succession of photos was taken (a spare set of camera batteries was essential for this day) and the train stopped some fifty yards past us. We scrambled alongside and climbed back on board. It had been possible to take photos virtually head on from really close to the track, the crew being totally unfazed by our antics.

Next stop was just outside the village of Krzycko Wielkie, where there was a small overgrown goods yard, closed many years previously. The old rails were virtually invisible beneath the long grass but the derelict office which had served it was reasonably intact, making an evocative setting for photography. Our loco stopped short of the office so that we could scramble forward, cameras at the ready. While we made our way to our chosen spot, several rounds of coal were put on the fire and a satisfying vertical column of black smoke headed skywards. We raised our arms to show that we were ready; the loco whistled, and the train stormed towards us, its three headlights blazing in the dull, late morning light. As it passed us, Andrzej shut off steam and the train coasted to a halt with a gentle squeal of brakes some fifty or so yards further on, before backing up to repeat the whole process.

After the second run, we scrambled aboard and continued for another mile or so, before spotting an isolated and dilapidated cottage by the lineside, served only by a dirt track. Alongside it was a pile of roughly-cut logs and, on the other side of the track, a colour-light signal with a large 'X' across it to show that it was out of use. Seeing it, I pointed and asked '*Proszę photostop?*' Andrzej nodded, '*Tak, tak, ok*', adroitly lighting up another cigarette with one hand whilst applying the brakes with the other. We drew to a halt in a cloud of steam and climbed down again. This time, as we had not really given much warning, Andrzej only backed up a short distance but the volcano effect was just as dramatic as our Pt47 and its string of vans rattled past.

Once more, we climbed aboard and took our places behind the crew, on the lookout for our next photostop. '*One more. OK?*' Andrzej warned us, raising one finger in case we could not understand and tapping his watch to remind us that time was of the essence. We travelled only a hundred yards before entering a forest. Spotting a dirt track that crossed the line, complete with a few road signs, I pointed it out to Andrzej and asked him politely, again in Polish, for our last photostop. Whatever he thought of my limited linguistic attempts I will never know but he never laughed at them, just as we never made fun of his English. He was one of several Wolsztyn footplatemen who realised that being able to speak some English would give them opportunities to make a few złotys out of visitors.

Before leaving the cab, I asked Andrzej if it would be possible to do two runpasts so that we could photograph the train from either side of the line. He looked at his watch and scratched his head. For a moment I thought he was going to demand a few złotys before he agreed in a pseudo-resigned but good-humoured manner. Upon leaving the cab, I winked at Andrzej and said, '*Proszę Big Big Smoke?*'. Thumbs up from him confirmed his understanding. We climbed down from the cab and the train reversed down the line, coming to a halt wreathed in steam, so much so that Andrzej had to wait for it to clear before opening the regulator.

A brief toot on the whistle, followed by loud barks from the dustbin lid-sized chimney, and the train rapidly gathered speed, roaring past us before stopping with a more dramatic squeal of brakes than previously. We guessed that Andrzej was trying to make a point as the loco was quickly wound into reverse, the whistle blown, and the train reversed at speed. We stayed on our side of the line until the train passed us, then scrambled across to the other side. For the final time, Andrzej gave the loco its head, once the clouds of steam had dispersed, and he had given us the obligatory warning whistle. Pt47-112 and its train of vans raced past us probably faster than on any of the other runpasts we had witnessed. Once the train had come to a halt with an even more dramatic screech of brakes, there was another toot on the whistle and the train reversed back to us. From the fireman's seat, Marek urged us to hurry so we clambered aboard as quickly as we could. I had scarcely got my foot on the bottom step before we were in motion. As we trundled towards Leszno, we had time to reflect on what had been a simply wonderful experience which we would have gladly repeated the following day if we had the opportunity. Sadly, we did not. This was the last time we were to witness a steam-hauled freight, albeit not a genuine one. From this year onwards, whilst the occasional freight was sighted in Wolsztyn, it was always diesel-hauled.

The Backhead of an Ol49

The timeless appeal of steam! A young mother watches with her child at Poznań Dębiec station whilst we wait for the 'Ice Cream Cart' to clear the section so that we can cross over to the Wolsztyn line

5

LIFE IN THE CAB

A HOSTILE ENVIRONMENT

A steam locomotive footplate is a hostile environment, fraught with potential danger. You have to remember that you are at the business end of some very heavy machinery, which can inflict serious injury or worse, if you are not constantly aware of your surroundings. I had already had a painful introduction as a teenager to the possibilities of injury in my native Wolverhampton. One evening in early March 1967, several of us teenagers begged a short ride on the footplate of a Black Five, which had worked one of the last steam-hauled trains into Low Level station. The loco pulled away from its train and the driver closed the cab doors. I leaned on them just as we turned up towards the Walsall line, prior to running back through the soon-to be-closed station. As we hit a short incline, the change of angle between the loco and tender trapped my left foot between the cab door and the tender footplate for just for a few seconds. The pain was excruciating, and I had a vivid bruise for a week or so afterwards.

By the time I arrived in Wolsztyn, I had already managed several memorable footplate rides on the National Coal Board after steam died on British Railways, being allowed to drive on one of them, so I was not a complete novice, but with Polish crews speaking little English, we had to look after our own safety. It was rare for the crew to give much advance warning of any danger; they expected us to be vigilant. All of the locos we rode upon carried warning signs – grubby, rectangular, yellow plates which read:

UWAGA! Wysokie napiece w siecie elektrycnej
zblizenie sie grozi smiercia

With two diagonal red flashes, no translation was necessary!

A BEWILDERING ARRAY OF CONTROLS

On the backhead, or 'firebox front' as some prefer to call it, there is a bewildering array of pipes, gauges, handwheels and levers. The blower is probably the most important control. It has to be opened each time the regulator is closed, and closed each time the regulator is opened. Failure to do this could cause the fire to 'blow back' into the cab. I have seen a momentary occurrence of this very scary and potentially life-threatening event. In front of the driver is a large, vertically mounted circular wheel, painted red – the reverser. In front of the reverser is a scale which shows the amount of cut off which has been set, the cut off being the amount of steam which is admitted to the cylinders. It is often simplistically compared to the gearbox of a road vehicle but is much more variable.

There are two brake handles mounted to the right of the driver – the nearer one for use when hauling a train; and the other one, mounted higher up, is the independent loco brake for running light engine. On top of the backhead is a large casting with four or five handwheels, and pipes travelling to many different corners of the cab. This is the manifold; it distributes steam straight from the boiler to all the controls which need it. Of the handful of gauges, one displays the boiler pressure and another shows the steam chest pressure: '*one maximum*', as we were constantly reminded. Most important of all is the water gauge, which shows the level of water in the boiler. If the level drops too low, serious problems can result. Almost as important is the gauge which indicates the pressure in the air-braking system. All around the cab there are pipes, pipes and more pipes of different diameters, leading to a multitude of different fittings. Surprisingly, there was no speedometer on any of the steam locos upon which we travelled! Cab doors and windows are often held in place by lengths of string or wire, which might fasten them in place but certainly fail to stop them rattling. Janus once pointed to a rusty hole in the cab front through which a gale howled: '*air conditioning,*' he joked.

I felt that I had a modest understanding of the mechanics of steam locomotives, how they work and how to drive and fire them. In theory … Putting theory into practice and understanding what the gauges were telling me was another matter altogether, exacerbated by my lack of the linguistic skills to ask questions. The crews were unable to share their knowledge and experience as much as I am sure some of them, at least, would have liked. It was one of the drawbacks of being in a country with a different language; there was so much that you could not question, or explain what you already knew. It must have been equally frustrating for the crews.

'VISSEL!'

The lines to and from Wolsztyn traverse mainly rural and agricultural areas and are crossed by many ungated and uncontrolled crossings. Before each crossing there is a warning sign to sound the whistle. With so many other things going through your head whilst driving, it was easy to miss one. The crew never did, hence the frequent cries of '*Vissel!*' The inner child in each of us thoroughly enjoyed sounding the whistle with great enthusiasm; I mastered the long, two-tone whistle that made us sound like an American loco crossing the prairies, but most crews frowned upon that! One year, the Authorities decided that motorists were so unused to the sound of a steam train whistle that they would ignore it, so the normal steam whistle was replaced by a diesel horn! Bizarre and surreal, this 'experiment' mercifully lasted only one season.

One year, my long-standing friend Bob, and I, witnessed first-hand an example of the abuse of these crossings, one which could have had fatal consequences. Not long after leaving Grodzisk, we spotted two cars following each other towards a crossing fitted with half-barriers. A long blast was sounded on our whistle. Both cars stopped. The first driver decided that he *could* make it to the other side, started to move, then almost instantly changed his mind when he realised that his gamble was too risky. The second car followed, but the driver's reactions were too slow to respond to the change of heart of the driver in front. Result: the second car ploughed into the back of the first one, shunting it perilously close to the track and giving Bob, who was driving at the time, a dreadful fright. We saw no more as our train proceeded towards Poznań but there were collective sighs of relief in the cab. Janus told us that one of the crews had returned to the depot after a night run from Poznań to find a substantial chunk of Volkswagen Golf hanging off the front of their Ol49. The loco was undamaged and they had not felt a thing as they drove on through the night, oblivious to the collision!

SIGNALLING

Besides being aware of what is happening with the loco, the crew have to keep a very close eye on the world outside the cab; there are signals and rules to be obeyed. The Polish crews dealt with such matters and I must admit that I never fully understood how the lines from Wolsztyn were controlled – there was certainly no token system such as exists in the UK. There are two signal boxes in Wolsztyn itself, one at each end of the station; these control the passage of trains by means of typical Continental-style semaphore signals. There are signals at some of the stations where trains cross; these are operated by the Stationmaster (who could just as easily be female.) Some of these stations have a nearby level crossing which the Stationmaster hand-cranks open and closed. Alongside the stations, there are large, cumbersome looking balance weights which tension the signal and point cables.

Once a train is under the control of signal boxes on the main line, colour-light signals have to be obeyed. This side of the running of the trains was entirely in the hands of the Polish crews. We had been told that control of the loco was to be returned to the 'real' crew whenever we operated under signals but this practice was honoured more in the breach than the observance and I have several times driven every inch of a journey from Wolsztyn to Leszno or Poznań. Being allowed to do this was a sign that the Polish crew had full confidence and trust in you; this had to be earned. They had to know that you understood how everything worked and that you would carry out any commands instantly. Whilst you were driving, the 'proper' crew might appear to be relaxing or fooling about playing practical jokes, but in practice they were watching you, the track and the loco like hawks. This was demonstrated on more than one occasion when, without hesitation, they jumped in and took control.

PREPARATIONS

A typical shift on the footplate would firstly involve putting on footplate clothes, usually beginning with a pair of denim jeans which had seen better days and which could be jettisoned at the end of our week. Engineman's coats were optional – only Henryk ever wore one. Bob wore his Jaguar work overalls, which really impressed the Polish crews – Andrjez even made him an offer for them! After the first couple of years, we settled on check shirts like the Polish crews. What we didn't copy from them was the typical bushy moustache! Gathering all the necessary equipment such as safety gloves, a bottle of water, a small snack and a camera and putting them in an old rucksack, we would set off either to the depot or to the station. Travelling on foot to the depot was pretty straightforward; on several occasions during our earlier visits, we saw farmers trotting along the road on horse and cart, oblivious to the cars and lorries which overtook them. At the end of the road to the depot were two shops which were, in effect, the front rooms of houses; one was a '*komputery*', self-evident as to its purpose, and one was a takeaway. Or so I thought until, upon investigation, I discovered that '*fryzjer*' was actually Polish for ladies' hairdresser ...

After risking life and limb to cross a busy main road, we would pass that Continental phenomenon the 'kiosk', which sells an amazingly wide range of goods, especially cigarettes, confectionery, bottles or cans of drink, lottery tickets, newspapers and magazines. To the left was 'Bar Bliźniak', mostly frequented by railwaymen having a drink or two after their shift, its frontage adorned by a large silhouette of a steam locomotive. We would cross the railway at a level crossing with full-width gated barriers, adjacent to the south signal box.

The depot was immediately beyond the level crossing. A hint of smoke would be noted as soon as the main road had been negotiated; there was always that thrill of anticipation at what the shift would have in store for us. The shedmaster, or '*Dyspozytor*', had his office by the entrance. Outside, was a green garden bench, often occupied by a couple of railwaymen putting the world to rights. In theory, the shedmaster would have some paperwork giving us official permission to be on the footplate but, in practice, this was never needed. Only once in ten years on the footplate did we even catch sight of any of these papers when, to our absolute astonishment, we had to sign one each. I wish I could have kept mine. The reason for this one-off occurrence became clear when a PKP inspector joined us in the cab at Tłoki, accompanying us to Poznań and back to Wolsztyn. Did he ask to see our paperwork? Of course not!

Our loco would normally be either on the main exit road or by the coaling stage. We would cross the turntable and wander up to it to introduce ourselves to the crew with a brief '*Dzień dobry, Jeff*' and '*Dzień dobry, Julian/Bob*'. Some crews would shake hands, others would scarcely acknowledge our presence, pointedly getting on with their tasks of preparing for the run ahead. We would climb up into the cab and stow our bags in the lockers on the front of the tender, fitting them in around the various sized spanners and other tools which filled the filthy, rusty, oily space. After a quick peep into the firebox and a check of the gauges, notably the steam pressure and the water gauge, we would await our crew. Occasionally, there would be a cry from below of '*Water on*' or '*Coal to fire*'. A few minutes before shed departure time, the crew would climb up into the cab and the driver would whistle for the signal to leave the depot. After a short, tender-first run down to the station – with much creaking, groaning, squealing and protesting from our loco – we would back onto our coaches. These would be coupled up to our tender by the fireman.

At this point, the crew would invariably climb down to the platform and sit on a nearby bench for a smoke until just before departure time. Occasional ribald remarks would be exchanged with passers-by and other railway workers, particularly if there was a female guard, before the crew climbed back onto the footplate. The guard would come alongside our engine, pass the time of day with the crew and hand over some paperwork. On more than one occasion, we have seen Andrzej simply screw this paper up and toss it straight into the firebox! Punctually, on every single run bar one out of the fifty or so that we made, our engine would whistle, the regulator would be opened and, with two deafening jets of steam from the draincocks, we would make our departure. By the time we passed the depot, our loco would be getting into its stride and the cut-off would be wound back from 60% to 50%.

THE EARLY SHIFT

For the first couple of years of our Experiences, the early train left for Poznań at 05:17. This meant setting the alarm for 04:00. The alarm clock in the room which I usually had was in the shape of an American steam loco, the actual alarm sounding like an American hooter. This would give plenty of time to come round, get dressed, perform one's ablutions, have the cup of tea which is surely obligatory for every Englishman abroad, and devour a bowl of cereal. We would leave the Society House quietly at about 04:45 and make our way down to the depot where our loco would be waiting for us, sizzling gently in the dark on the exit road, control of which was by colour-light signal. The south-end signal box is directly opposite the depot so the signalman could see when we were ready to leave. Occasionally, we would find that the loco had already left the shed, having gone down to the station to couple up to its train in order to pre-heat it for passengers, in which case a swift march along the tracks ensued. On a normal morning, up to fifty people would board our coaches. By the time we reached Poznań, the train would be packed full, many passengers standing – or sitting on the floor.

In other years, departure time varied between 05:05, 05:27, 05:32 and 05:56. In 2004, the time was brought forward to 04:16! Fortunately, this even more unearthly hour lasted just one year. Perhaps there were no passengers at that time of day? A few shovelsful of coal would be placed around the firebox and the injector put on if necessary. At departure time, the guard would wave his or her flag, the 'official' fireman would shout '*Odjacz*', one of the crew would acknowledge with a wave, the regulator would be opened and, with a token 'toot' on the whistle, we would be under way. Leaving the platform end, our loco would lurch almost immediately to the left and then, just as immediately, to the right, its coaches swaying along behind us. Light from the coach windows spilled across the track, illuminating sleepers and ballast. The atmosphere in the cab was just wonderful, the driver silhouetted against the steam and smoke passing his window. Looking back out of the cab, a trail of smoke would hang behind us, streaming along the carriage roofs and beyond. On cooler mornings, steam would issue from beneath the coaches from the steam heating.

Our loco would shake, rattle, roll and thump its way over points and joints. Maintaining balance was a constant battle. Comfortable it was not; elemental it most certainly was. Behind the cab, the tender would have a life of its own; as the loco rose out of a dip, so the tender would drop into it. You might be tempted to lean against the tender front, but you would only try that once! Every now and then, as the tender hit a bigger dip than usual, lumps of coal would fall through the coal hole, bounce off the shovelling plate and roll around the cab floor. The air in the cab would be alive with a mixture of smoke, steam and tiny particles of coal dust dancing in the swirling draughts. If you were to blow your nose, your handkerchief or tissue would turn black. It made you realise what an efficient nasal filtration system we possess. And make you wonder just what *was* going into your lungs ...

Just to add to the general mayhem and organised chaos on the footplate, there would be many gaps in the floorboards and between loco and tender. Perhaps most alarming of all for the faint-hearted are the massive gaps above the trailing pony truck wheels, which are within touching distance should you be so foolish. In brief moments of respite from your labours, the rotating wheels could be watched. Beneath your feet, the cab would buck and plunge at each rail joint. At some joints there would be a definite lurch; keeping your footing would be a delicate balancing act. At the controls, the driver would wind back the cut-off every hundred yards or so and maybe just nudge the regulator open a touch more as the loco found its feet. Passing the depot, and then a factory, the line veers off to the right and begins to climb quite sharply in a long, left-hand curve, taking it over the line to Leszno. At this point the driver would usher whichever one of us was taking the first drive – usually Julian or Bob – into his seat and the fireman would vacate his seat. It would not be long before *'Water on'* would rouse me from my brief reverie, nerves jangling with the dread of fluffing my first use of the injector. As a very general rule, if I got it right first time then the rest of the run would be fine. If I failed ...

To drive the loco from a standing start, the reverser would be wound out to 60% on a marked scale, using the red wheel. This took some considerable physical effort and was easier for us to do standing up; the Polish crews had the strength to drive sitting down. A small peg locks the reverser mechanism in place and the regulator is opened by pushing a large lever to the left. On some classes of loco this would be easier than on others but if the regulator was opened too fast, the loco would slip. (*'Piwo!'*) To cure this, it would have to be closed very smartly, then gently reopened as the slipping subsided. Roughly every hundred yards or so, the reverser would be pulled back in 10% steps as far as 30%.

A brief dash through fields and past darkened houses would take us to the first stop at Tłoki, where illumination could best be described as 'feeble'. A lineside marker would give 300 metres warning of a station, so the regulator would be closed some distance away, the blower opened and the reverser wound out to 60%. Braking would begin at the time the driver gave the instruction; a little 'brake' would be put in and the handle nudged back to its starting position. If judged correctly, the train would stop at the right place, marked by a trackside cross. If judged incorrectly, it might be necessary to put in a little more braking or even to 'dump the air' if it looked like the train might come to a halt too soon. I was always wary of braking and never really felt that I had mastered it. For this reason, I found that I preferred to fire rather than drive. Both Julian and Bob were much better at braking than I was.

The 'coal to fire' gesture would be made as the regulator was closed and the blower opened. I would place some twenty rounds in the firebox, six shovelsful to the left, six to the right, two to the front and half a dozen in the back corners and under the door. The net result of this effort would be dense, acrid clouds of smoke rolling down from the chimney, sliding lazily past coach windows and in through any open ones. As the train drew to a halt, any waiting passengers would be enveloped in a smog of chokingly thick, sulphurous, black smoke. I always wondered what the poor 'customers' thought of their daily kippering!

In order to see where you were going, it was better to put your head out of the side window, thus exposing you to all the swirling steam and smoke. It was quite normal for small particles of unburnt coal to dance in the exhaust, so wearing glasses was a boon, unless it was raining. Goggles would be an obvious choice to protect your eyes but would probably have been seen as somewhat less than macho. Hanging on to your precious PKP cap was strongly advisable, especially with Marek waiting to pounce and sell you a new one! On the early shift, the air could be quite cold so you would end up with sore eyes and skin. However, trying to keep a lookout through the grubby front cab window was almost impossible because the great length of the boiler severely restricted your field of view and, in any case, the glass was rarely clean. The same procedure would be followed at the rest of the stations on the route – places with unpronounceable names but whose platforms would be busy, even at such an early hour (I have no recollection, however, of even a single passenger joining us or alighting at Drzymałowo, the scene of our exploit with a firework). At Grodzisk, the controls would be relinquished to the 'official' driver as we entered the station and until we had departed. This was not always the case, particularly if 'Mad' Andrzej or Janus was in command.

Arrival at either Poznań or Leszno would give us the dilemma of whether to stay on the station or to accompany the crew to the depot and hang around like spare parts whilst the loco was watered, oiled round, generally checked over and turned – if it could be accommodated on the turntable. Generally speaking, at Poznań we would stay at the station to watch the fascinating comings and goings, but at Leszno we'd accompany the crew to the depot, which was more interesting to wander around.

The road approaches to Poznań station had food and drink stalls with lower prices than those on the station; the *'kantor'* here gave better currency exchange rates too. As previously recounted, my budding Polish came up short the first time I confidently ordered *'dwoje hamburgeri'* – which turned out to be a double-decked hamburger, overflowing with salad and mayonnaise, not the two hamburgers I had expected. I had to order another one for myself … they were so huge that Bob and I struggled to finish them. It was here too that the two of us had our first encounter with proper Polish coffee, in a very small cafe beneath the bridge across the station. I was quite proud of my *'Proszę dwa kawa'* – until I attempted to drink the stuff. It turned out to be a 50:50 mix of thick grounds and hot water …

THE LONG LAYOVER

For timetabling reasons unknown, in 2005 the Leszno working involved a very long layover at the city's depot. The train would leave Wolsztyn at 05:56 for a journey of about an hour, but return from Leszno would not be until 15:28, the crew staying with the loco. When Andrzej was involved, this gave him time for even more practical jokes, invariably at our expense. On one of the layovers, he managed to start a large fire in the grass around the loco; putting it out involved lots of stamping and dancing. Another time, we indulged in an England versus Poland football match in the cab, using the open firebox doors as a goal, the football being empty cans or balls of screwed up paper. Obviously, when a goal was scored, the ball could not be retrieved so another had to be sought. Once or twice we even used a piece of coal! At this stage, the fire would have been allowed to settle down to just a thin glowing bed of coal.

The Polish crews had food which had been warming up on the steam manifold for most of the journey, usually in the form of sausages wrapped in tinfoil and eaten on a roll. On one occasion, one of the crew heated a can of soup on the front of the boiler. Forgetting to pierce the can resulted in a messy explosion! Whilst Polish crews didn't have billycans of tea warming on a shelf above the firehole as they did in Britain, there were occasionally bottles of what certainly did not look like cola.

A typical station – Tłoki, now a private home

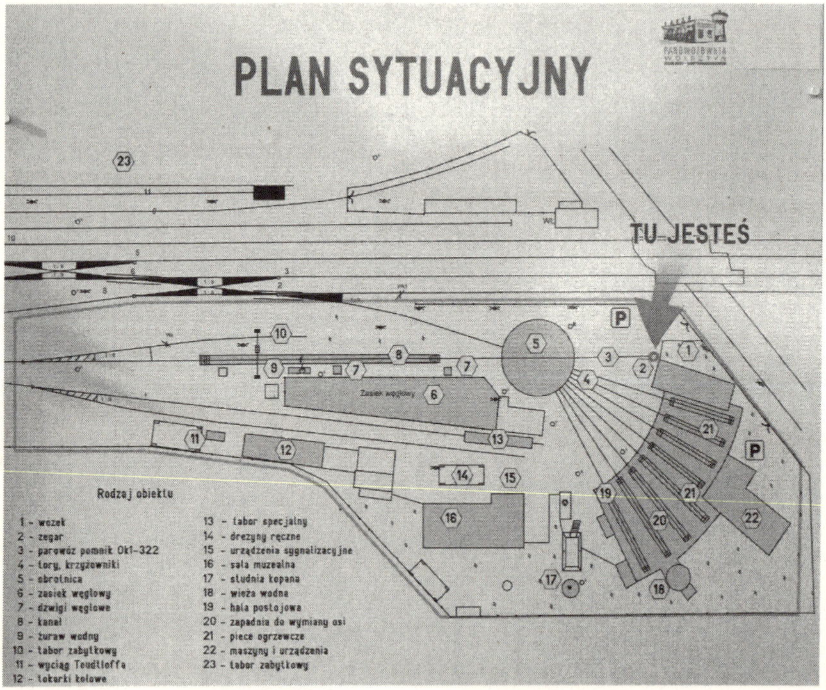

Plan of Wolsztyn Depot (C.R. Hatt)

6

PAROWOZOWNIA WOLSZTYN

THE REAL THING

Wolsztyn's steam motive power depot is a living, breathing, working museum. Unless you were lucky enough to experience first-hand the atmosphere and excitement of a working steam depot in the years before Britain's railways finished with steam, you might find it difficult to understand what they were really like and what a thrill it always was to visit one. Wolsztyn gives you an insight. Back in the days of steam on British Railways, the illicit nature of most visits took the adventure up a level and gave the enthusiast an extra shot of adrenaline. After all, which self-respecting steam-age trainspotter has *not* been chased out of a Motive Power Depot? To one who has 'been there', the Wolsztyn depot brings it all back in a vivid assault on the senses because it still is 'the real thing'. It is an introduction to what it was really like, minus the fear of being thrown out. Or worse!

Built in the late nineteenth and early twentieth centuries, Wolsztyn shed is a typical Continental-style half-roundhouse. It has eight covered roads, each with a pit. These roads are fed by a turntable which leads from the depot yard; several of the roads are semi-permanently occupied by locos which are out of service but kept under cover to minimise deterioration. By and large, it is kept fairly tidy but there are plenty of tools and components lurking about to trip the unwary (how do I know?). Inside the shed building are several large space heaters, resembling bug-eyed alien monsters; we never actually saw one in use since we never visited in the depths of winter. A small blacksmith's forge stands at the rear of one of the shed's eight roads; the accompanying anvil and work-table sit close by. Oxy-acetylene welding torches and their associated gas tanks are in regular use, not something I remember from the British sheds of my teenage years. Spare air pumps lie scattered around the depot on pallets. Given the number of problems these pumps seem to cause, this is hardly surprising and they are probably exchanged at fairly regular intervals. There are also great piles of brake blocks, the replacement of which is a matter of routine.

At the rear of the depot building are several well-equipped workshops with smaller machine tools, some ancient, some more modern, and a few small side rooms containing various pieces of equipment; one holds an old, ornate stove. The depot staff are capable of turning their hands to repairs and servicing of all kinds, from routine boiler washouts to tube cleaning. It is quite normal for valve gear parts to be removed from a loco for refurbishment 'in house' and firebox repairs too are commonly undertaken on site. Heavier jobs are sent to Leszno depot as it has bigger machinery available; Wolsztyn does not possess a wheel drop. In the past some major work went to Gniezno.

Adjacent to the depot, there is a building which was once an enginemen's hostel but is now used for accommodating visitors. It also contains a small museum, together with lockers for PKP staff; some of the lockers were crammed absolutely full of 'souvenirs'. As we had soon discovered, some enterprising footplatemen would attempt to sell them to unsuspecting victims. Including us!

At the heart of the depot is an electrically-operated turntable, controlled from a small hut at one end. This turntable featured in one of the most eye-popping sights witnessed in all of my visits to the depot but, unfortunately, I had no camera with me. Every year in Poland there is a National Children's Day, when pupils visit all kinds of workplaces. Most years, at least one class of local schoolchildren visits the depot, their behaviour impeccable as they roam freely amongst the piles of ashes and cinders. The culmination of their visit this particular year was to stand on the deck of the turntable whilst an Ol49 was turned through 180 degrees. The sight of several dozen small and excited schoolchildren, wreathed in steam and all lined up with school bags in hand, was simply unbelievable to one conditioned by the Health and Safety ethos of the UK.

For me, much of the memory of the age of steam lies in the all-pervading smell of steam, smoke and oil. This aroma is present in abundance but there is so much more to Wolsztyn, even down to the fact that, just like British sheds at the end of steam, there are many withdrawn locos scattered around, slowly rusting away. Many of them were abandoned there in the 1980s and 1990s; some have been in the same place on every visit, some seem to have moved every year. In the early years of our visits, Ty51-223, a huge 2-10-0, stood near the coaling stage, having already been there for decades, but by 2007 it had been replaced by another 2-10-0, this time a slightly smaller one – Ty45-379 – which Julian and I had driven and fired in 2002. By 2009, this loco had in turn been displaced and moved into the station yard, along with several other rusting hulks, none of which could have the remotest chance of ever steaming again. Its place was taken by Pt47-112, a loco very familiar to us.

Throughout every single visit, yet another 2-10-0, a three-cylindered, Prussian designed Ty1-76, minus some of its valve gear, has stood at the end of the road opposite the coaling stage, a large mound of ash and clinker effectively barricading it in place. It was repainted in 2012 but by 2018 the weather was beginning to take its toll and it looked little better than when we had first seen it. Stored in the goods yard on the opposite side of the track, there are several rows of locomotives which will surely never see service again. Some have been cosmetically restored and look quite presentable, others resemble nothing less than scrapyard wrecks, their platework rusting through like lacework, among them being several of the world-famous German Kriegsloks with their 'bathtub' tenders. Built in their thousands during the Second World War, some are in good external condition whilst some look only fit for scrap but we never saw one in traffic. The former goods yard next to these locos is used for would-be drivers to practise their skills in anything from a small hatchback to a bus or even an articulated lorry! Also in this yard is an office-workshop which, I think, is the equivalent of an MOT station.

In more recent years, some of the locos in the worst condition have been moved near to the station. Surprisingly, and unlike what would almost certainly happen in many other countries, we never saw any of the locals climbing or playing on the wrecks, which are more or less free of graffiti.

COAL AND WATER

Locos are coaled from a great stockpile of the black stuff, presumably of Polish origin. Looking back over a dozen years of visits to the depot, I never once witnessed the arrival of a consignment of coal so I am unaware whether it arrived by road or by rail. I presume it was the latter, unloaded by the depot's rail-mounted grab crane. Occasionally, locos are coaled by this crane, which grabs a bucketful of the fuel from the stockpile then travels along the track to deposit its load in the tender. Most times, however, coaling is done by means of a conveyor belt, which loads coal from the depths of the stockpile into small tubs. These are then lifted above the tender by a small crane – the loco crew knocks a catch on the tub and it spills its load into the tender.

Taking on water is usually carried out simultaneously with coaling and ashpan cleaning. Rather than having a small 'manhole cover' opening into the water space, as in the UK, most Polish tenders have a long, hinged plate on each side of the coal space. With one of these open, there is no need for the tender to be placed with any great accuracy next to the water crane.

When a loco arrives back on shed after its day's work, its own crew, together with a member of shed staff, carry out disposal duties. The fire is thoroughly cleaned and raked through until much of it has dropped into the ashpan and thence into the pit below. Water jets are used to hose out the ashpan but the crew will usually climb down into the pit and do some raking of the ash from the ashpan too. It is a very dirty and dusty business but the crews take it in their stride.

That the shed staff were very dedicated and very capable was vividly demonstrated in 2005, on the day of our 'fake' freight train photographic outing. We rode back in the service train behind Pt47 112 rather than in the cab as we had hoped. The two facts may or may not be connected but the 'apprentice' driver was very heavy handed with the regulator, making the loco slip many times. When it arrived back on shed, a broken driving wheel spring was discovered. After the fire had been cleaned, no time was wasted in cannibalising a spring from withdrawn Pt47-106, which conveniently happened to be inside the roundhouse. The broken one from our engine was removed and replaced by one off the withdrawn one, all done whilst the loco was in light steam, yet it was ready in plenty of time for its duties the next day. Replacing the spring would certainly have involved at least two men working well into the evening, in conditions which could not be described as pleasant.

Occasionally, a 'pupil' would be asked to assist with the rocking of the grate from within the cab but usually they were confined to cleaning the white-edged motion and wheel rims. We quickly discovered that part of the Wolsztyn Experience ritual was to clean these at the end of a loco's working day. Doing this could be quite intimidating as jets of steam could and would issue forth from many different places. A special supply of old rags was normally kept in one of the lockers on the tender; most crews would hand over a selection of rags at the end of the run, often without saying a word. They knew that you knew what they wanted; we felt it important to 'muck in'. Cleaning the valve gear and wheel rims in this way left the crew to do the more important tasks involved in disposal of the loco at the end of its shift. Polish steam crews are very much jacks of all trades; they not only drive and fire the locos but help out in servicing them, and lend a hand with any kind of repair work, whether it be emergency or planned.

By day, there is plenty to see when a loco arrives for servicing or when it emerges from the roundhouse for its day's work. Clouds of steam surround the locomotive as it moves about the shed yard; when it stops, the crew fuss over it, oiling every moving part and constantly checking every component. In this way, in 2005, Andrzej found Ol49-7 with a tyre which had shifted on a driving wheel. Repairs were done at Leszno; within a week it was back in Wolsztyn and back in traffic. By night, the depot takes on a magical aura and is a photographer's paradise. Any loco in steam will make ticking and hissing noises all of its own; wisps of steam issue from all kinds of places. If the firebox door is open, the whole cab takes on an orange glow and any person on the footplate or on the ground will take on a ghostly appearance. Many of my best depot photographs were taken at night.

The depot is at the heart of the railway community and it is quite normal to find locomen on their rest days, just sitting around chatting. Just across the track from the depot is Bar Bliźniak, which we knew as the Enginemen's Bar. After a good day out with a friendly crew, some course participants would invite them to this bar for a drink or two, all the time trying to remember the old adage, *'Never try to out-drink a Pole ...'*

We never tired of spending time at the depot and in 2019, three of us visited the shed. Russ had never been there before; and his reaction was priceless. We loved exploring and reminiscing about our teenage exploits in the UK, and for Russ it brought back childhood memories that he had once thought impossible to relive. Wolsztyn does that to you.

Parowozownia Wolsztyn: a very special place indeed ...

A cigarette for Andrjez and a beer for Janus.
Unknown member of shed staff watches on

Ol49-23 on the Exit/Entry Road at Wolsztyn

Ty1-76 and Ty51-223 dumped at the depot

The shed crane coals Ol49-69

7

POLISH STEAM LOCOMOTIVES… ARE BIG!

POLISH STEAM LOCOS are much larger than British ones – as indeed are diesel and electric ones – thanks to a much more generous loading gauge. Photographs of the visit of Great Western Small Prairie 5521 to Poland in 2007 show vividly the massive difference in height and width. At the time of our visits, this little gem was locked away in a shed at Wrocław, so we never saw it.

The majority of the locomotives which work in and around Wolsztyn are very workmanlike and simple to maintain, with labour-saving features such as rocking grates which can be raked and hosed out at the end of a working day. American influence is evident in the use of bar frames for the more modern locomotives. Sandboxes, and they are big ones which could easily be mistaken for a dome, are mounted on the top of the boiler, just as in the USA or other Continental countries, resulting in a mass of pipework cascading down to track level. In reality, I was never aware of sand being applied to the rails; the trains we hauled were never heavy enough to require its use.

The air pump, a constant source of trouble on some of our journeys, is situated on the running plate on the left-hand side of the smokebox. Usually on the right-hand running plate, close to the cab, can be found the steam powered turbo-generator which provides current for the huge headlights, of which there are normally three on the front of the loco – two on the bufferbeam, the other on the smokebox door. Three are also fitted on the rear of the tender, as well as other lights in the cab. High running plates are the order of the day, leaving all of the wheels and motion exposed. It may not be the most aesthetically pleasing of features but shows a very pragmatic approach. I doubt whether anyone could describe Polish locos as 'pretty'; they are very much functional and practical. Poland is a country of few railway enthusiasts and I would not imagine that they have ever argued over the relative aesthetic merits of a Pt47 over an Ol49!

Partly for geo-political reasons – Wolsztyn was once under Prussian rule – there has been a massive German influence on Polish locomotive design. Just like their German counterparts, locomotives in Poland have a penchant for red-painted wheels, valve gear and frames. Wheel rims are painted white, as are the raised edges of the red-painted valve gear.

The most common colour for locos is black; no lining is involved. Occasionally, the cab will be in green, as will the cylinder covers. Items like cab window frames and the sides of the running plate will be in red, as will handrails. This colour might also be randomly applied to smokebox door hinges, ladders on the rear of tenders and other parts of the loco. Some locos, notably the Ol49s, are fitted with miniature snowploughs and these too are painted red, often outlined in white. Pm36-2, the big Pacific, was painted in an emerald green when we first visited Wolsztyn but it, along with Pt47-65, is now in a very dark green, which can look almost black in some lights. Looking a little odd to British eyes – though there were exceptions – smoke deflectors are usually painted the same colour as the rest of the locomotive.

THE LOCOMOTIVES WE DROVE AND FIRED

Ol49: The Ol49 2-6-2s are the locos which we drove and fired the most. The 'Black Fives' of Polish steam, they are simple, reliable and willing. They steam freely and, fortunately, are tolerant of rookie firemen such as myself. Well over half of the footplate turns we had from Wolsztyn were on this class. With their sprawling pipework and smoke deflectors mounted next to the chimney, they are far from elegant but I grew to love them; they were the easiest to drive of all the Wolsztyn locos we encountered. We rode on them in various states of repair, drove one which only had one working injector – and a temperamental one at that – and have even ridden on one which had a boiler leak the equivalent of someone leaving on the tap in the kitchen sink. But they still did the job.

Pt47: Next in line were the two Pt47 2-8-2s upon which we worked. These locos were built in the post-war period and represented the finale of Polish passenger locos. They probably have the cleanest and simplest of lines of all Polish locos and are relatively uncluttered by external pipework. Their large smoke deflectors are noticeably angled in at the top towards the smokebox. Not the easiest of locos to work on, mainly due to the sheer size of the firebox, their great power was never taxed whilst hauling a handful of coaches, but the grate still had to be kept covered. Being mounted in a very long firebox, this was no mean feat! When driving them, the sheer amount of power available meant that a slip was inevitable if the regulator was opened too far or too fast. There are three Pt47s at Wolsztyn but only Pt47-65 remains in service for special workings. I have driven and fired 112 but both it and 106 are unlikely ever to steam again. Both of these locos have been cosmetically restored and the last time I saw them they both resided in the yard opposite the depot, 106 being in glossy black.

Pm36-2: For the first few years of our visits to Wolsztyn, the lone Pacific, Pm36-2 'Beautiful Helena', was the star turn, but in 2008 we found 'her' laid up outside Leszno depot, having burst a main steam-pipe whilst hauling a special train. Built in 1936, and one of only two class members, the first one was built with a streamlined casing. By 2010, the Pacific had been repaired and was back in traffic, painted in a very dark green. We did not have all that much experience of the 4-6-2, but it certainly gave a smoother ride than any of the other locos we rode upon, and had an incredibly spacious cab.

Ok22-31: For several years from 2004, Ok22-31 was in action, mostly on specials but also working the timetabled

Poznań services with great success. Although the first few of this class of 4-6-0s were constructed in Germany, the vast majority were Polish built. This loco and its classmates were not fitted with smoke deflectors, unlike almost all of the other classes which we encountered, and had lower running plates than the more modern designs. It was a dream to drive and fire, even the injectors seemed simple. The first time I saw it was amongst the 'abandoned' locos parked in the old goods yard opposite the depot but it was restored in 2004. Sadly, by 2019, it had joined a line of derelict engines near the station, a very forlorn sight.

Tr5-65: Built in Berlin in 1921 by Orenstein and Koppel, this attractive little 2-8-0 was originally constructed as an 0-8-0. It was in occasional use on schoolchildren's specials; I only ever had one chance to work on it, when I drove it on a very special working from Wolsztyn, the first leg of a journey to Wrocław for a few days starring in the filming of a Polish movie. Like Ok22-31, this loco also lacked smoke deflectors, or 'blinkers' as we used to call them in the UK.

Ty45-379: The loco of my painful, one-off experience with a spinning reverser wheel, Ty45-379 was a post-war heavy freight design of 2-10-0. Many of this class were built in Poznań by Cegielski, a firm founded in the 19th Century by Hipolit Cegielski.

One of the shed staff cleans out the ashpan of an Ol49

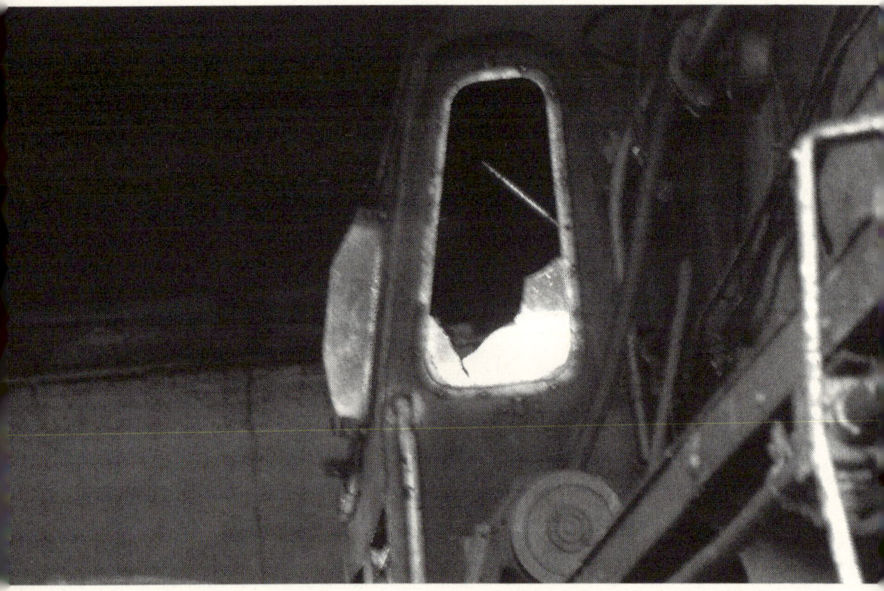

Bob's Broken Cab Window

8

THE (SLIGHTLY) SCARY SIDE AND THE (MUCH) LIGHTER SIDE

CONCRETE

In 2004, Bob and I had a full week on the footplate between Wolsztyn and Poznań, even enjoying an extra turn on Pt47-65 on the final Saturday, supposedly our day off. We had bought tickets to ride on the 11:20 steam-hauled Saturdays and Sundays-only Poznań train, but Andrzej spotted us on the platform. Of course he had other ideas, waving away all protestations of being unsuitably dressed, too tired or having no gloves.

'Why you no in cab with Andrzej?'

'It's our day off-holiday.'

'Holiday? Holiday? What is holiday? You drive!'

'But we have no gloves.'

'Is no problem, I find.'

Little persuasion was needed; I drove as far as Grodzisk then took over the shovel from Bob, who drove the rest of the way to Poznań. He also drove all the way back, a most exhilarating, fast run which saw us having to wait time at several passing places. Andrzej's partner on the footplate was Piotr, the driver who had seemed so unfriendly two years earlier. He was definitely lightening up with the passage of time and left me to do the firing with minimum intervention. And the injector worked sweetly!

This was the year of the early departure at 04:16, necessitating a 03:00 start. We drove it on the Thursday, reunited with Zenon and Henryk. The former was still totally relaxed about life on the footplate, the latter still permanently on edge. We also had to contend with difficulties with the water gauge, making it difficult to assess how much water was in the boiler, more of a problem for the Polish crew but it still affected us as it put them on edge. Somewhere around Ptaszkowo, whilst Bob was driving, and as we passed beneath an overbridge, there was an almighty crash. A lump of concrete smashed the front window a couple of feet in front of Bob, who instinctively reached for the brake handle. Immediately realising what had happened, Henryk stopped him from applying the brakes (presumably, this was not the first such occurrence). Henryk offered to take over but Bob was quite happy to stay in the driving seat, still thoroughly enjoying his somewhat draughty drive. Arriving punctually at Wolsztyn, Bob was not deemed worthy of the besom treatment!

INTO THE LIGHT

The same year as our encounter with the concrete block, Bob and I worked a late turn from Poznań. As we rounded a bend near Drzymałowo, a couple of bright headlights appeared on the track in front of us, about a mile ahead. Bob alerted Henryk, the official driver, to the situation. Henryk immediately told Bob to apply the brakes gently whilst he both sounded the whistle and flashed our own headlights several times. We slowed down to virtually walking pace, a palpable feeling of anticipation, bordering almost on fear, gripped the four of us in the cab, even though we knew that what we were seeing was impossible. The fear lessened gradually as we realised that the distant lights were stationary as we rolled slowly forward towards them. The darkness in the cab, relieved only by slivers of orange light escaping from the firebox, added to the tension. Time seemed to stand still.

The whole incident lasted no longer than two minutes. As we crept up to the lights, it became clear that what we were seeing was the headlights of a car which had parked on a bank overlooking a curve on the line, pointing down the track towards us to aid a photographer, who was filming our train. The photographer in question was well known to the crews, a Japanese enthusiast who visited Wolsztyn at least once a year and who had a reputation for appearing in dangerous positions very close to the track. Our crew fired off a volley of *curvas* and other Polish profanities at the unrepentant photographer, along with a lengthy and very angry blast on the whistle. Whilst the footplate crews were fairly tolerant of railway enthusiasts, they definitely resented those who came along merely to observe and contributed no money to the cause of steam, especially those who got too close to the rails or shone car headlights down the track! Some of the more enterprising footplatemen, however, found ways of getting money out of the linesiders, as will be related shortly …

In this instance, the photographer's poor reputation was very well founded. The following day, we very nearly killed him. Foolishly, he ran across in front of our train just as we started off from a small wayside station. Marek saw him just in time, pushed in front of Bob and did an emergency brake stop, bringing us to a juddering halt. We thought, but could never be sure, that one of the loco buffers had actually caught him. Needless to say, he received a whole broadside of *curvas* from both of the crew, followed up by a tirade from the guard, who was not amused at being jolted to the floor as she climbed back into the train after giving the Right Away.

LOOK – NO BRAKES!

On the Wednesday of my 2006 Experience, I worked on my own on the 08:47 to Leszno with a crew who had best remain anonymous, for reasons which will become apparent. I had fired all the way to Leszno and was driving back to Wolsztyn. All went well until, nearing Boszkowo, I applied the brakes. Nothing happened. I assumed that I was doing something wrong but the driver ushered me urgently out of my seat and stopped the train right at the very end of the platform by putting the loco into reverse. The crew both climbed out of the cab and attempted to fix the air pump, but after ten minutes of hammering and spannering, gave up trying to effect a repair.

While the crew attacked the pump, I had time to notice a hut near the lineside which was graffitied with 'Punk-Rock Forever!' in English. After a brief discussion and a muttered '*Pump kaput*', they decided to carry on to Wolsztyn.

The driver offered to direct me whilst I continued to drive – '*Nema problema, I tell you*' – but I declined and insisted that he drove back whilst I fired instead. Things were taken much more slowly than usual, each stop being made by winding the loco into reverse and speed not being allowed to rise to anywhere near its normal level. Arrival at our destination was some half an hour late and the loco was immediately taken out of traffic for repair. Technically, the crew should probably have waited for a rescue loco but this would have made the train even later, delayed passengers and brought repercussions further down the line. I doubt whether the reason for the train's late arrival was booked as anything more specific than 'loco problem' but that evening, the offending machine was on shed in light steam, its broken pump removed and a replacement fitted. All in a day's work!

A QUICK FIX

On the Thursday of our 2011 Experience, both Julian and I had a poor night's sleep thanks to another course participant snoring so loudly that he could be heard through two thick walls. We had walked to the shed to work on Ol49-7 on the 06:28 to Leszno with Mad Andrzej and Irek – our Dream Team. The trip had been eventful thanks to the usual tricks from Andrzej: pieces of coal on the driver's seat, tying Julian's jacket to the tender front (whilst he was wearing it, of course) and had been excellent fun all round. I drove all the way from Wolsztyn to Leszno, apart from the first few hundred yards. As we were leaving Wlosakowice, Andrzej, the official fireman, suddenly barked, '*Jeff, brake!*'

I applied the brakes. Hard. We stopped rapidly from

around 20 mph. Our eagle-eyed not-so-madman had spotted the left-hand lubricator linkage breaking off. He climbed down from the cab, retrieved it and then used his mobile to contact the depot at Leszno to order repairs. Before restarting, the crew poured oil on all the left-hand surfaces which were normally lubricated mechanically, an operation which they repeated at each stop. On arrival at Leszno, I was allowed to drive right up to the turntable before Irek reversed us into the shed for repairs. Miraculously, a spare lubricator arm appeared, probably having been removed from one of the derelict hulks lying around the shed yard. With Andrzej supervising and some of the depot staff assisting, the replacement arm was fitted with remarkable speed. We eventually left Leszno only half an hour late, Julian driving all the way to Zbąszynek, running round and buffering up before I drove the return working tender-first to Wolsztyn. Whilst not especially scary, this experience showed us again that no matter how light-hearted the atmosphere in the cab, the crew were on full alert every second of the journey.

TROUSERS

Most of the lighter moments in the cab seem to have involved Mad Andrzej, an absolute born comedian but also engineman extraordinaire. Slightly overweight, like so many of the crews, he would chain-smoke his way through a turn in the cab whilst simultaneously running through his catalogue of practical jokes. When not clowning around, Andrzej would spend much of his time texting or talking on his mobile. I have more than once seen him drive into Poznań with one hand on the brake handle and one texting on his battered Nokia.

Not a practical joke, but on one of our runs in 2006 we worked on Ol49-23 with the 05:16 to Poznań. The official driver was 'Spit', or Stanisław, to give him his proper name, with an unusually subdued Andrzej as fireman. (Hangover?) I fired all the way to Poznań with no assistance from Andrzej who just let me get on with the job. Julian drove, with plenty of 'advice' from Stanisław, who on another occasion was to refer to my own driving skills in very uncomplimentary terms. We stayed on the loco to the depot and had an excellent Polish breakfast in the depot canteen. It was here that it was pointed out to Andrzej that he had a great rent in the seat of his trousers. He disappeared, returning moments later with borrowed needle and thread, then nonchalantly proceeded to remove his trousers in full public view in order to repair them. The sight of Andrzej in his boxers with needle in hand and thread between his teeth is not one to be easily forgotten, which leads neatly on to one particular year when a youth on the outskirts of Poznań took to 'mooning' the daily steam workings. Obviously a discerning mooner and one who was 'rewarded' with long and loud whistles from the crew and the loco. Marks for artistic merit?

Having repaired his trousers and restored his dignity, Andrzej drove us back to Poznań station where we hooked onto a set of four vintage four-wheeled coaches. These had been retrieved by an ancient looking diesel shunting locomotive from the depot, where they had been in temporary storage. He and Stanisław then led us on a circuitous tour around the Poznań area, ending up at the outer suburb of Kiekrz, where our loco ran round its train and took us back tender-first by a different, but equally circuitous, route to our starting point. We then caught one of the new units (known disparagingly by the steam crews as 'Ice Cream Carts') back to Wolsztyn. Just like modern units on Britain's railways, this one boasted extremely uncomfortable seats which felt as if they had no padding in them whatsoever.

BRIBERY AND CORRUPTION PART TWO

Throughout much of our week in 2011, a party of German steam enthusiasts was staying in the Wolsztyn area, popping up all over the place with cameras and video cameras at the ready. They would follow trains from station to station by road, badgering the crews with their constant requests for lots of smoke, and driving at high speed between stops. Never ones to miss financial opportunities, both Marek and Andrzej had allowed themselves to be bribed to provide the photographers with 'Big Smoke'. Marek would even go so far as to put extra coal on the fire for the 'special effects.'

I doubt whether the Polish crews spoke much German so conversation between the two nationalities was always conducted in English; for once we had no problems in understanding what was happening. Besides a minimum of 20 złotys per 'Big Smoke', Marek would also accept beer or 'wodka' as currency, whilst Andrzej, surprisingly, preferred hard cash rather than 'wodka' – or cigarettes at a push. On one run with Marek that week, the Germans took it in turns to ride in pairs in the cab from one station to the next, the price for this being fifty złotys per ride! When we arrived back in Wolsztyn, several bottles of 'wodka' mysteriously appeared on the footplate. Probably from Netto …

RACE TO LESZNO

In 2011, we did a Sunday afternoon turn to Leszno with Ol49-59. We had watched it return from Leszno in the morning, then went to the shed at 13:00 to pick up our loco, which was to haul just two of the old double-decker coaches. With a crew of another Andrzej – this one generally referred to as 'Not-Mad' Andrzej – and Stanisław, we had a good run, Julian driving out to Leszno whilst I both fired and had a very a good success rate with the injector, in spite of Andrzej insisting on adjusting it by a millimetre every time I opened it. On the return journey, when I was driving, he did the same with the regulator!

As we rounded the bend into Leszno, our first colour-light signal was against us and we pulled up alongside an 'Ice Cream Cart' which was stopped at an adjacent signal. Simultaneously, both signals turned green and there then followed a 'race' into the station. Many of the passengers on both trains entered into the spirit of the occasion with much highly vocal cheering. Some of the doors on our train were opened and passengers leaned out of the carriages yelling encouragement to our driver. Naturally, the drivers of both trains engaged in a battle of the whistles. The 'race' lasted about two minutes before we entered the station in triumph, 'winning' by a loco length. Goodness only knows what the signalmen made of our race but, as the signal box at Leszno looks straight down the line, I would not be in the least bit surprised if the whole thing was a very enjoyable set up!

STRING

Brake problems seem to have been a recurring theme during our various Experiences. One year, when driving with Bob, there was a problem releasing the train brakes due to a faulty valve on the tender. Remarkably, the ever-resourceful Andrzej had solved this problem with the aid of a strong but tatty piece of string, which had to be pulled before the train set off. Several times he forgot. The loco and tender moved but the coaches refused to budge. What the passengers thought of the constant jolting can only be guessed at but the guard left us in no doubt about her feelings!

Air Pump (C.R.Hatt)

Px48-1919

The terminus at Śmigiel

Narrow-gauge rural idyll. A runpast at Śnieciska on the Środa District Railway

9

ON THE NARROW-GAUGE

ŚMIGIEL

I must admit that I have no great affinity with narrow-gauge railways, even though I have enjoyed riding on most of The Great Trains of Wales as well as the Lynton and Barnstaple Railway. I expected narrow-gauge railways in Poland to be much less polished, certainly not tourist-centred, and much more down-to-earth than their Welsh and English counterparts. I was not wrong but I had to wait until 2008, and my seventh Wolsztyn Experience, to be offered the chance to ride on the narrow-gauge.

The opportunity partly came about because of problems with our planned experience. No standard-gauge loco was in a serviceable condition. Our substitute experience was to be on the narrow-gauge line from the village of Śmigiel to the small town of Stare Bojanowo – which was on the main line between Poznań and Leszno. It would be 'interesting' to say the least!

Our day out began with an hour-long car ride, four course participants being chauffeured through deeply traditional farmland to Śmigiel by Witold, the dapper and well-dressed Wolsztyn shedmaster, in his Nissan Sunny. Julian and I enjoyed the company of Joe and Jay Monty, father-and-son Americans from Massachusetts. They were very knowledgeable and experienced footplate men, both quietly spoken, who volunteered on a line in their home state in the USA. Joe, an engineer by profession, had the advantage over us by way of sporting a Polish style moustache. Witold spoke a moderate amount of English, so on the way to the railway he did his best to give us a little of the history of the line which we would be visiting.

The line opened in 1900; originally built to metre gauge, it was regauged to 750mm by PKP following the Second World War and would carry upwards of a quarter of a million passengers a year. It was also heavily used by freight trains, mainly bringing coal into the area, the line extending as far as Rakoniewice on the Wolsztyn-Poznań line. By the turn of the century, with traffic reducing quite dramatically, the line had been cut back to the section between Stare Bojanowo and Śmigiel. By 2008, the remaining line was on its knees and trains virtually non-existent. At the time of our visit, traffic was confined to services carrying children to their schools – normally operated by diesel railcars – with steam only operating for Wolsztyn Experience participants. In previous years, standard-gauge wagons had been carried piggyback style on narrow-gauge transporter wagons, known as *'Rollwagen'*, feeding a transhipment yard at Stare. This traffic had come to an end several years before our visit, although one standard-gauge wagon on its transporter was still to be seen near the loco and stock sheds.

As we neared Śmigiel, Witold pointed out to us the remains of narrow-gauge track running along verges beside the road and across people's driveways. From time to time, the remains of goods yards could also be seen, some with trackwork still in situ but no rolling stock in sight.

It was a slightly misty morning as we entered the station yard at Śmigiel, and we were greeted by the sight of some waggons, coaches and locomotives in a siding. Most of the coaches were in very good external condition and had obviously once been used for tourist services, having the upper halves open to the elements. At the end of the siding was an out-of-traffic 0-8-0 tender locomotive, Px48-1765; next to it stood an out-of-service narrow-gauge diesel of the Romanian-built Lxd2 class, its numberplate removed. The steam engine had a little indecipherable graffiti scrawled across its tender and cabside. Much of the track around the station was weed-strewn and overgrown but most of the rails had a light polish on them. Witold parked next to the well-maintained station building, which sported a large sign proclaiming in Polish '106 Years of Railways in Śmigiel.' Carrying with him the 'manbag', which many Continental men sport, he led us across the tracks and introduced us to Krystow, our driver for the day.

There were four low platforms, all very uneven and covered in weeds to match the track, and a goods shed combined with a booking office. Standing in one of the island platforms, and in steam, was another Px48, this one numbered 1919, looking clean and well cared for. A plate on its dome revealed that it was built in Chrzanów in 1951, so it was younger than I! Coupled to the tender-first loco was a trailer that had clearly seen better days, its red and cream paintwork faded to a matt finish. One side of the trailer was heavily graffitied, the other side completely free of any 'artwork'. Completing the picture was the station cat, a tabby which quickly adopted Julian. Or was it the other way round?

The locos and rolling stock all sported large, central, curved buffing plates and were coupled together by two small screw couplings, one either side of the buffing plate and linked together by a pair of centrally pivoted levers. Our not-so-little loco had a two-man crew; its driver, Krystow, was very welcoming and friendly and spoke more than a little English (a lot more than my Polish!) and there was also a fireman whose name I never learned. As the footplate was relatively small, it was only possible for one of us to fit in alongside the crew at any one time, so it was necessary to take turns in the cab, the other three course participants riding in the trailer car. We let Joe take first turn on the 'throttle', as he called it, on our first run to Stare Bojanowo.

OLD-FASHIONED THRUPENNY BITS

Leaving Śmigiel, we passed the locomotive depot/works on one side and a stock shed on the other. A diesel locomotive was parked near the depot and a couple of trailer cars, identical to ours, stood outside the stock shed. Beyond those two buildings there were sidings containing a few transhipment wagons, all obviously out of use. Once out of the station confines, the line curved to the left and dropped sharply away into pleasant countryside.

Our trailer was spartan on the inside, sporting wooden bus-type seats, but in reasonably presentable condition. It had two windows at each end, flanking a central opening door. One of the end windows displayed a paper sign proclaiming 'Stare Bojanowo'. These windows gave us excellent close-up views of the loco hauling us; we could watch the footplate crew in action. At one end was a small compartment with a primitive wood burning stove for heating the coach, there being no provision for steam heat from the loco. Opposite that, in another small compartment, was a toilet. Examining its contents, we realised that we would have to be pretty desperate to use it, but it did bring back happy memories of toilets on British Railways carriages, where you could see the track when you flushed!

The line was undulating and meandering with scarcely a yard of straight or level track. We travelled alongside rough dirt tracks and past fields full of sunflowers, passing very close to farmyards. Many of the curves were more akin to the old-fashioned, twelve-sided thrupenny bits of our childhoods; progress round them was by a succession of lurches. Joe prosaically described it:

> *'Basically, two streaks of rust sticking up from the mud and weeds.'*

Riding in the trailer car with its bogies, we were more cushioned from the poor trackwork but there was no such comfort in the loco cab as the Px48 class do not have any form of guiding truck. Naturally, speed was low, probably no more than 15 miles per hour maximum; the poor nature of the track would have made anything much higher suicidal!

After a twenty-minute journey, we made our way into Stare Bojanowo station. Our entrance was permitted by colour-light signal, with the last few hundred yards being quite steep and tortuous. Just before curving sharply uphill into our terminus platform, a line branched off to the right and dived under the main line, to end in a goods yard/transhipment station. With our final destination in sight, we crossed a main road at a level crossing equipped with barriers and warning lights, before pulling up next to the main station building. Our platform was screened off from the main platforms by a sagging and rusty wire fence. A few yards short of the end of the line, our train came to a gentle halt. Krystow uncoupled our Px48 and ran forward to a short headshunt, scarcely long enough for our loco, before setting back over a crossover to run round the trailer. Negotiating the crossover was done ultra cautiously, at very low speed, and we were asked to watch that the wheels of our loco remained on the track. This involved standing between the rails, bending down and watching the loco inch towards us, quite hair-raising! Once across the points, Px48 1919 chugged off round the bend which we had traversed to enter the station. Another cautious negotiation of a pair of points, aided by the use of a crowbar, and our 0-8-0 reversed back to its trailer. It then pushed the trailer back all the way to the end of the line, before coupling up to it.

After a short break, we set off back to Śmigiel, now with Jay at the controls. This time it was possible to observe events on the footplate from behind the tender, which had sticks of kindling wood on top of the piled-up coal. Very little effort was required from our loco and it seemed to do a lot of coasting, thick black smoke trailing behind it. From the other end of the trailer, it was possible to look back at the track which we had passed over, marvelling at how we had managed to stay on it with so many missing or rotten sleepers!

Back at Śmigiel, there was time for a break and to fill the tender of Px48-1919. The blue-painted tap for the water

Above: Leaving Ruchocice, Zenon attacks the air pump on Pm36-2

Right: The dreaded injector

View from the driver's seat. All marked in red, the large wheel is the reverser, the large casting right is the train brake, with the loco brake above, used only when running light engine. The lever at bottom right dumps the air from the braking system in the event of braking too hard

Viewed across the turntable, locos bask in the sun at Wolsztyn depot

A shed worker ghosts alongside Ol49-23 during servicing

Night-time splendour! Ol49-59 rests beside the coaling stage

On the narrow-gauge at Stare Bojanowo, awaiting the arrival of schoolchildren

In early morning sunshine, the 05:05 from Wolsztyn to Poznań leaves Grodzisk and its tall signal box

At Poznań station, the author stands next to EP07-1003 which he has driven part of the way from Wrocław

Light and shade inside Leszno depot. Former 'Kriegslok' 2-10-0 on the left was once a stationary boiler; electric loco EU07-235 is under repair on the right

Marek oils round Ol49-59 as it is serviced at Poznań depot

Inside Franowo depot, Poznań, electric loco ET22-944 is raised up on jacks for repair

A grand entrance! Ol49-69 and Ol49-59 thunder into Wolsztyn station with empty coaching stock

To the accompaniment of waving flags, Pt47-65 leads the 'Pirat' through the streets of Kołobrzeg

Above: Journey's End. Photographed from the top of the lighthouse called Latarnia Morska, our LOTOS diesel has reached the end of the line at Kołobrzeg Port

Left: The author

column was below ground and required a grid to be removed before it could be turned on. This had to be done lying on the rough ground; there was no mad rush for this task so Julian was 'volunteered', his newly acquired friend, the station tabby, keeping watch over his efforts. After our break, and with me at the controls, we set off again for Stare Bojanowo.

Relatively little firing needed to be carried out, and Krystow did most of the driving, allowing each visitor only a little use of the regulator. Running precariously through farmland and pleasant countryside, we soon found ourselves back alongside the main line, leaving the trailer parked in its customary place. I stayed on the footplate and was allowed to drive the loco through its run-round procedure and up to its trailer. Misjudging the brakes – nothing new there – I hit the trailer with some force, pushing it several yards further up the platform. Krystow smiled ruefully and took over, placing the trailer gently into its rightful position at the end of the line.

THE SCHOOL BUS

It was explained to us that we now had a long layover so, after eating our packed lunch either in the trailer or, in my case, on the footplate, the four of us Experience participants went to explore the disused transhipment sidings and some of the wagons which still lay there rusting away. We marvelled at the sharpness of the curves and the appalling state of the track on the short line which led under the main line and up to the transhipment sidings. We heard several years later that the bridge beneath the main line was filled in so these sidings were finally put to rest.

Whilst we were exploring the transhipment yard, several expresses passed on the electrified main line, including one *INTERCITY* express hauled by blue Class 83 lookalike EU06-001; this class was nicknamed *'Szóstka'*, which means 'the sixth.' Unfortunately, the photograph I took of it was very blurred but as we waited, a formation of four cranes flew gracefully over our heads. By this time, Poland had been in the European Union for several years and Polish railways were beginning to follow the practice of most European railway systems of partial or total privatisation. This was demonstrated by an orange and black ET182 Co-Co electric owned by STK but leased to the Polish freight company LOTOS, which passed us on a long freight train.

Returning to our train in the narrow-gauge part of the station, we were informed that we were actually waiting for some real passengers! Shortly afterwards, about a dozen children appeared, all clutching satchels. The oldest among them could not have been more than ten years old, the youngest around six. There was no adult to accompany them but their behaviour was excellent. Once all of the children were on board, we set off at our usual leisurely pace on the final run to Śmigiel, Julian taking his turn at the regulator. Our progress was even slower than before as we stopped half a dozen times in random places to drop children off virtually at their front door. By this time, I had learned the Polish for 'Goodbye' – '*Do widzenia*' – and was rewarded by a succession of surprised smiles and '*Do widzenia*' in return. What a way to travel to and from school!

Back at Śmigiel, we said our goodbyes and thanked Krystow and his colleague before rejoining the waiting

Witold in his car for the journey back to Wolsztyn. I was not the only one who fell asleep in the back of the Nissan!

PAINT

Our second and final encounter with Śmigiel came in 2010, which was the 110th anniversary of the local system. An Englishman named Robert Hall had done a lot of work trying to revive the line, which was more or less closed – some might have said 'mothballed' – and was trying to drum up publicity. Some local politicking had saved the line temporarily but the reprieve was short-lived and it was to close for good the following year. Some work had been done on repainting one of the diesel railcars and a press photographer had been lined up to capture a few images of 'the mad Englishmen' at work on some of the restoration. Howard had asked Julian and me to take part in a photo session in the shed at Śmigiel and we had willingly agreed.

The day chosen just happened to be the day that we were rostered to work the 05:11 from Wolsztyn to Poznań. I had driven Pt47-65 all the way from Tłoki to Wiry under the strict eye of Henryk who, thankfully, had by this time desisted in his use of the birch twig besom for misdemeanours. Either that or I got everything right … I fired for much of the return journey but Henryk took over when he thought I was not man enough for the job because I could not get coal to the back of the huge firebox. It was a scorching hot day and I didn't argue! On arrival at Poznań, we had breakfasted on delicious liver and onions taken in the railwayman's café for a nominal price. After a shower and lunch, Julian and I were driven to Śmigiel by Alex, Howard's partner. Howard came along too; as noted before, he didn't drive and a veil should really be drawn over Alex's driving style …

At Śmigiel, the photographer – another Marek – joined us in the shed, along with Robert. We were duly photographed wielding paintbrushes, diligently applying red paint to the side of a railcar. In pristine blue paintwork in the same workshop, was one of the narrow-gauge diesels, pots of paint still standing on its running boards. People had obviously been busy around the place; there was hope for a bright future. It was very sad that all their efforts came to nought. In Poland, it seems that the local mayor has a lot of say over what happens to local railways and, politics the world over being what it is, different mayors wield a lot of power and can blow hot and cold on local matters. The steam services at Wolsztyn have been known to suffer a similar fate, with more than one hiatus in their operation.

This day though, once Marek was happy with his photographs – and we never did see the results of his endeavours – we were able to adjourn to a nearby bar for a beer, a bite to eat and *'lody'* – ice cream – to help us cool down on what was a very hot day, so hot in fact that most of us were wearing shorts.

After a leisurely snack, Howard suggested we go back to the railway and he would 'see what he could do'. Ten minutes later, a railway employee – possibly the only one – started up one of the railcars and we were off for a jaunt along the line. Not only that, but Julian was at the controls, taking instructions in Polish from the regular driver! It was interesting to see the line from the very front of the train but more than a little disconcerting to see close up, once more, the perilous state of the track that we were traversing.

At Stare Bojanowo, we went in search of more ice cream before I found myself in the driver's seat trying to make sense of the Polish instructions. In truth, driving the railcar was not difficult; the biggest problem was trying to keep the speed down to the prescribed 20 kph and, of course, I had to contend with constant leg-pulling from the passengers in the main body of the unit! The control panel had lots of very impressive switches, none of which were needed for our journey, but there were also several gauges. Nobody seemed to take much notice of those either; all that seemed to matter was the throttle and the brakes. Unlike the mainline electric and diesels that we had driven several days earlier, the throttle was in the shape of a lever rather than a large spoked wheel.

Back at Śmigiel, we had time to notice and photograph a very smart red railcar belonging to SKPL Regio parked outside the depot. Known as '*Lokomotywownia KD*', this was an attractive twin-tracked building with a tall chimney at the rear. Then it was time to say our goodbyes and leave for Wolsztyn, this time travelling in the photographer's battered green Ford Fiesta. As we left, a torrential thunderstorm erupted. It felt appropriate as we were never to see Śmigiel again.

ŚRODA

My final encounter with narrow-gauge came in 2019. On a very hot and sunny day, Bob, Russ and I made our way to the small town of Środa Wielkopolska, some 50 miles north east of Wolsztyn. Approaching our destination, we passed Howard with a larger-than-life American named John

Schmidt (known to all as 'Boomer') and his lanky friend Lovatt. Howard asked if we would give the Americans a lift to the narrow-gauge station as John was struggling in the heat, the three of them having made the journey by train to the mainline station at Środa. We quickly discovered that Boomer and Lovatt were great company, very generous, and with many tales to tell of railroading around the world. Unlike many enthusiasts that we met, they were also very good listeners.

We soon arrived at Środa Miasło station, located in a street named Mieczyława Niedziałkowskiego. As we parked, we thanked our Satnav for getting us there since none of us would have liked to ask a passer-by for directions to a street with that name! The line here had closed in 2001 but reopened in 2003 under the auspices of the local authorities, who ran it on and off at weekends as a tourist service, and still do. The trains were known locally as '*Chiuchia*' ('Choo choo') but the railway itself bore the official title of '*Średzka Kolej Powiatowa*' – Środa District Railway.

The first thing which caught our eye as we entered the terminus was a mock-up of Thomas the Tank Engine mounted on a flat bogie wagon. We were unsure whether or not to be impressed by this but decided that if it brought the line extra revenue which might enable it to stay open longer, then we were in favour. We were greeted by the news that there was also a diesel-hauled special train running for local schoolchildren because it was Children's Day. There were, therefore, no 'proper' coaches in which we could ride. Alternative arrangements had been made for our steam-hauled excursion and, with typical Polish invention, they were slightly 'off the wall' ... We had plenty of time to wander around the site, which was situated in an industrial area, surrounded by several factories and warehouses. Close to the depot/works, we noted a set of steam locomotive frames, complete with cylinders, and a small diesel locomotive in PKP colours.

Our train consisted of three bogie wagons, the first two of which were boxcars, the leading one of which was for the guard who, in a surprising fit of Health and Safety, sported a fluorescent green gilet emblazoned with the title '*Kierowanie Ruchem*'. We never did ask what that meant but his duties included stopping traffic at the many road crossings encountered on our excursion. The last wagon was an open, flat bogie wagon which had some flimsy 'safety' tape tied along the sides between its stanchions. Its effectiveness in preventing anyone from falling off the wagon was very dubious. And, by the way, there was no tape or barrier at the end of the wagon, beyond which was the track. But the *pièce de resistance,* wedged in place in a very rudimentary and precarious fashion, was a garden bench which had been placed on the decking of the wagon for us five travellers to sit upon! To be fair, it was wedged in place with some wooden blocks. At least it was to begin with!

Hauling our train, tender-first, was Px48-1756, another member of the relatively youthful 0-8-0 tender engines built in the 1950s for narrow-gauge lines. I had, of course, encountered the class before – at Śmigiel. In contrast to locomotives on the likes of the Welsh narrow-gauge, these machines looked quite modern and had a very purposeful appearance, with both loco and tender sporting huge headlights.

Leaving the terminus just after ten o'clock, we passed

a few houses and some large tanks containing molasses for the adjacent sugar factory before skirting some allotments, which are quite popular in Poland. Our unusual train ran alongside the main line from Poznań for the best part of a mile. We passed sidings full of standard-gauge oil tank wagons and a wayside halt called Środa Wask, where a local looked askance at our passing, calling out a question to our driver. We passed more allotments before climbing up and over some standard-gauge tracks in a manner reminiscent of the way the main line at Wolsztyn climbs over the Leszno line. The bridge here was allegedly life-expired, so the very existence of the line was threatened.

Continuing at a leisurely pace, and paralleling for a short distance the standard-gauge on one side and a country road on the other, we passed through pleasant, rolling fields and countryside before arriving at our first stop: Słupia Wielka. Here we detrained; the train ran back a hundred yards before doing a series of runpasts for the benefit of us passengers. All, that is, except Howard, who elected to stay in his seat on the garden bench for the runpasts, waving regally as he sailed by. The two Americans rode in the crowded cab with the young Polish crew, taking it in turns to drive. Three runpasts were done before the train stopped in the station for what we were told would be a lengthy wait for the schoolchildren's special to cross, so our loco and wagons sat in one platform, simmering gently in the warm Polish sunshine. Nobody seemed to know when the diesel would arrive. Nobody seemed to care. We were assured that our train would wait for us so we set off the short distance to the village shop where we used our (very) limited Polish to buy ice creams, cold drinks and snacks.

Returning to the station, which had a tiny waiting room made from an old van body with one side removed, and which had recently been painted bright green, we didn't have too long to wait for the schoolchildren's special to arrive, hauled by a smart red diesel loco. The guard of this train must have been extra safety conscious as he sported an orange fluorescent top with green bands around it and the same two words as our own guard. After a ten-minute break, mainly for the crew of the diesel to have a chat with Howard, both trains set off in opposite directions, our guard shepherding us across the road next to the station. Once across the road, the train stopped for him to climb aboard his makeshift brake van.

A mile or two further on, we stopped at Śnieciska, which had a very small corrugated-iron waiting room, vaguely resembling an old air raid shelter, and which was sweltering in the heat of the day. Here, the Ceremony of the Runpasts was again performed three times, thus enabling us to take photographs of the train from several different angles. A surprising piece of lineside furniture was a concrete table tennis table! Once again, our train was watched across the road adjacent to the station by our guard, holding up a rather small, yellow, triangular flag as his symbol of authority. Here, there was a small concrete bridge over a stream and several poppies growing by the roadside, making a picturesque touch of variety for some of our shots.

A loud whistle from the Px48 summoned us back to the luxury of our garden bench. We could relax until we reached the terminus at the small town of Zaniemyśl. Our fourteen kilometre journey took us a leisurely hour and forty minutes, every one extremely enjoyable and relaxing, a reminder of what it must have been like on country branch lines in the halcyon days before World War Two.

DONUTS. WITH VODKA

Zaniemyśl station yard was spacious with what had once been a goods yard, all contributing to an impression of what a busy line it must once have been. Our loco was detached from the train, running forward along the only platform to take water, then parking itself in the bright afternoon sun for a rest. The terminus was right in the heart of town, so we went for a brief stroll into the very pleasant town centre. The sun continued to shine down on us and Howard took us into a small bakery where he treated us to coffee and an absolutely delicious vodka doughnut each. From the taste of the doughnuts, it was obvious that liberal amounts of vodka had been used in their baking!

After strolling back to the terminus, we found that, in our absence, our train had been re-formed. Our flat wagon had been shunted round to the back but this time it had a barrier at the rear in the form of a raised timber construction. We also found a pair of Italian railway enthusiasts engaging the loco crew in conversation, their hire car parked in the former goods yard. We were to see more of them shortly.

Just over an hour after our arrival, we reluctantly left the picturesque town of Zaniemyśl and began our return journey. What had been a really good day turned into an outstanding one as our train, now with our loco chimney-first, pulled up again at Śnieciska. Here, we were joined by the two Italians, together with a member of that rare breed, the serious Polish railway enthusiast, and a rather overweight, Lycra-clad Pole on a racing bike, looking as if he were fresh from the Tour de France. There was much surreptitious mirth at this particular sight.

Three more runpasts were done; on one of them, Lovatt kindly gave up his seat so that Russ could have a turn. This time we had to contend with the four other camera-toting enthusiasts but it was all very good natured. At one point, one of the photographers shouted that someone was in his shot leaning on a pole. '*It's not a Pole, it's an Italian!*' I shouted and the whole party descended into fits of uncontrollable laughter. Once our three runpasts were over, we all re-boarded our train, all that is except the two Poles and the two Italians who set off at great speed for our final runpasts at Słupia Wielka. When we arrived there, the Poles and the Italians had easily beaten us to it. With the engine now running chimney-first, it was possible to obtain some different photos from the morning session, especially some 'light and shade' effects, as the sun had moved round.

All good things come to an end. After our final runpast, we set off back to Środa but the fun was not over yet. As we left Słupia Wielka, a country lane ran parallel to the line, on our right-hand side. The four Polish and Italian enthusiasts had managed to get ahead of our train again and were both photographing and videoing us as we ran alongside them. Not only that but the Italian driver was videoing us. Whilst driving. On the wrong side of the road!

It felt a little anticlimactic as our train made its way back to Środa, paralleling once more the main line, where a signalman in his elevated signal box watched our passing with apparent amusement – or was it amazement? Finally, we passed the sugar factory, where we noted a diminutive standard-gauge shunter manoeuvring some open bogie wagons into the yard. Some four and a half hours after leaving Środa Miasło, we were back at the terminus after what had proved to be one of the most enjoyable days any of us has ever had on a railway line anywhere in the world, topped off by dining that evening at 'Pizzeria Mafia' in Wolsztyn!

Our luxury coach, with garden bench, on the Środa narrow-gauge line.
(C.R. Hatt)

Thomas (Tomás?) gets everywhere!

Ol49-7 leaving Rakoniewice

Leaving Strykowo

10

2009 – AN EVEN MORE MEMORABLE YEAR

BY STEAM TO WOLSZTYN

The time of year of our Wolsztyn visits differed; the earliest was the beginning of April, the latest, the end of September. The majority of trips were around the end of May/beginning of June, when the days were long, so it was possible to photograph the early morning departure in daylight. Doing so depended on being awake early enough to walk down to the level crossing by the depot, then walk up the Leszno line far enough to see the train swinging away on the Poznań line. Being awake early enough depended on what train or trains had been worked on the previous day, or would be worked on the day in question. The weather was rarely an issue, and the cooler air of five in the morning gave excellent smoke and steam effects.

In 2009, I made the journey to Wolsztyn on my own, flying from Liverpool to Poznań for the first time. My plane taxied to within a hundred yards of the terminal building where the passengers boarded a brand-new bus, which took us on a lengthy grand tour of the airport before depositing us by the passenger entrance. It would have been quicker to walk …

At Poznań station, there was a small, low-key exhibition of modern locos, along with both passenger and freight rolling stock. Notable among the exhibits was a brand-new electric loco of a very modern Euro design, sporting a new PKP Cargo livery. This was probably the first loco I saw in Poland which carried both its class number of Eu43-009 and its Union Internationale des Chemins de Fer (UIC) number of 627 005-7 2151, not exactly a memorable one!

I waited several hours for the 17:12 steam-hauled Wolsztyn train, the ticket for which cost me just over £2 for 50 miles. There were three Americans in the cab of faithful old Ol49-59, one of them videoing proceedings. I told them that I too was there for the Experience and we fell into a lengthy conversation, only cut short when the crew of Henryk and Piotr demanded that their 'students' actually did something other than chat! Arriving at the Society House, I discovered that one of the other course participants was Simon, a South African who had much more steam experience than I, especially on the African continent, but also at Wolsztyn. I had met him briefly before but had never worked with him. He was an experienced and knowledgeable engineman with a deep love of steam.

A MAJOR FIRST AND BRIBERY AND CORRUPTION PART THREE

The Tuesday of this year was a notable one indeed. It marked the very first occasion that I was ever allowed to drive every single inch of the journey from Wolsztyn to Poznań, this being on the 05:05. Simon was fireman and the official driver was Andrzej, who was in an especially benevolent mood and more or less left me to my own devices. Only at Grodzisk did he put his head out of the cab window to make it look as if he was in charge. It is very hard to put into words the feeling I had as our train came to a halt at our destination. A sense of real achievement and a mixture of elation and relief probably best sums it up!

I was all set to fire the whole way back to Wolsztyn but, just before departure, we were joined on the footplate by an eccentric young Pole, who was obviously known to the crew and whose presence was clearly tolerated by them. He insisted on doing most of the firing but when I persuaded him to take a rest and let me do some of the work, he immediately began telling me in his fractured English how, where and when to put coal in the firebox. Much to my relief though, he was quite happy to operate the injector for the whole journey and I was only too happy to let him!

On the Wednesday, I fired for an Englishman called Peter who was doing the 'Experience' as a birthday present for his 40th birthday. His family was in the train behind us and his young daughter Karina joined us on the footplate for some of the run; I photographed her in the driving seat as we sped between stations. She was not wearing what might be considered remotely suitable clothing but nobody, least of all Andrzej, seemed to mind … apart from 'Mr No Happy'. He was singularly unimpressed by having to share his cab with so many others, muttering *curvas* to himself the whole time. His mood was not improved as the cab became even more crowded when an ageing German enthusiast climbed aboard at a wayside station and sat on the dickie seat behind the driver for much of the journey. Firing round the legs of all these passengers was not easy! I'm not sure how much the German rider had paid Andrzej but he was joined by a succession of his compatriots, who were riding in the train behind us and would ride in the cab, in pairs, for one stop for a mere 20 złotys. Each!

LINESIDING

Simon had a hire car, which he invited me to share on several occasions; he also knew his way round the area, having scouted some good locations for photos during a previous visit. Following our first shift together, and after a shower, a rest and lunch, we went to Granowo Nowotomyskie to photograph Ol49-59 on the 13:31 from Wolsztyn to Poznań, soon after it left the station. Simon found a dirt track through a field suitable for photography and for us to wave to fellow novices in the cab. Before we could do that, we had to wait for the diesel unit travelling in the opposite direction to pass us, Granowo being a timetabled passing place. There could hardly have been a greater contrast between the brand-new SA132 and the old Ol49 accelerating noisily through the fields. Simon then drove us on to Stęszew; our road ran parallel to the line, with the train travelling a frustrating hundred yards away across a field from us. At Stęszew, the train was held again to cross with another unit so we were able to photograph it leaving the lengthy and dilapidated platform, with a derelict water tower in the background.

Returning to Wolsztyn for a cup of tea, we later departed for Rakoniewice, where we set up in a field about a third of a mile from the station, to await the 17:12 from Poznań. We managed to get several different shots of it, including one with a church in the background and one side-on shot of the loco as it passed on the embankment just above us. Noticeable in almost all of the lineside photos is the presence of that phenomenon which seems almost to have vanished on British railways: telegraph poles and wires.

Two days later, we hit the jackpot as neither Simon nor I had a footplate turn. I set my alarm and arose early on a beautiful sunny morning, walking up the Leszno line a quarter of a mile or so to a vantage point opposite the old furniture factory to watch Ol49-7 leave on the 05:05. 'Mr No Happy' leaned grimly out of the fireman's cab window, obviously determined to hang onto his perch as long as possible before deferring to a mere trainee. After feasting on these five minutes of action, it was back to bed for a few hours and a leisurely morning before joining Simon on the brief drive out of Wolsztyn, as far as the bridge over the Leszno line to watch the 13:30 departure. From our vantage point there, it was possible to trace the trail of smoke and steam from Ol49-59 as it left the station and before it could even be heard. This remarkable loco was nearing the end of a marathon one hundred and fifty-two days in steam, and was more than overdue a boiler washout. It chattered its way past us, and onwards towards its first stop at Tłoki, its rhythmic exhaust notes lingering in the air for several minutes. The excitement of watching this spectacle properly from the lineside for the first time was so tiring that I went back to the Society House for a siesta.

Our next foray was to a forest track close to Ruchocice, where we didn't have to wait long before the 17:12 from Poznań bustled past us doing about 50 mph. Ol49-7 was at the head of three olive-green coaches, leaving behind a glorious trail of smoke and steam. From Ruchocice, it was but a short drive to Rakoniewice to set up again in the same field as the previous day, this time slightly nearer to the station, to attempt even better shots of the actual departure. The photos turned out no better, but it was still awe-inspiring just to watch and listen to the sights and sounds of the last daily, timetabled, steam-hauled passenger service in Europe, if not the world. We tried videoing and photographing simultaneously and the driver gave us an obliging whistle as the train drew near. To our chagrin, a local farmer chose that exact moment to drive his ancient and noisy blue Ursus tractor right past us! However, there was still time to get back to Wolsztyn station to watch the loco shunting its coaches into the correct platform, ready for the following morning's 05:05 departure.

The next day – Friday – would be a long one. After photographing and videoing the departure of the 05:05, this time from nearer to the signal box, Simon had his car ready to make a dash for our first stop, in the forest near Ruchocice, on a different forest track from before. We had scarcely climbed out of the car when we heard the loco whistle for a distant crossing, followed shortly afterwards by another whistle for the crossing that we had chosen. Emerging from behind a stand of trees, the early morning light caught the exhaust of Ol49-7 as it accelerated fiercely away from its last stop. There was white steam aplenty from the loco; the fireman's side injector was on and the escaping steam from it trailed around the tender and under the leading carriage of the three-coach train. Twin headlights glowed brightly in the early morning light. (For some unexplained reason, this loco only had two front headlights.) As the train passed us, it rounded a bend, the early morning sunlight flickering through the loco's exhaust, steam trailing from the rear coach's steam heating.

Magical though this spectacle was, we had no time to lose as we sped off towards our next objective: Grodzisk. Parking in ul Dworcowa (Station Road), we made our way across the running lines to a vantage point in the middle of some disused sidings, close to what had once been a small locomotive depot. Also in the yard, but of little interest to us at the time, was a rusty white Fiat 500 – the original type, not the modern reincarnation – mounted on railway wheels! We didn't have to wait long before Ol49-7 appeared, stopping in the station close to the platform-mounted signal box. By this time, the sun had risen higher in the sky, the moon was faintly visible and the light was right behind us, making for some beautifully atmospheric photographs as the train departed, steaming vigorously on its way past the former goods shed and a rail-served factory. A rake of PKP blue, open, bogie coal wagons was parked alongside a stockpile of the black diamonds visible near the track. There were at least two people videoing the departure from the train itself, and one other man in the station yard besides Simon and myself. It was one of those truly memorable steam moments as the train clattered away past the station's smaller signal box on the next leg of its journey to Poznań.

Leaving Grodzisk in the Poznań direction, the line describes a huge arc, passing the town's football team's stadium, amongst other buildings, and then heading towards the next stop of Ptaszkowo. This gave us plenty of time to get ahead of the train again and reach the village of Strykowo. Here, on yet another dirt track through the sandy-soiled forest, we managed to find a viewpoint which was head on to the station building. Shortly afterwards, Ol49-7 pulled in, exactly on time. Departure from Strykowo was positively volcanic, with clouds of black smoke shooting straight and high into the still, clear morning sky. As the steam-hauled trains were on a tight timetable, they were always worked very hard away from each stop. If you were in the driving seat, the Polish driver would urge you on if you didn't give the loco its head smartly enough.

We stood close to the track for as long as we dared, stepping back only on hearing the loco whistle at us, cameras clicking away in the early morning sunshine. We moved away and stood in some bushes to watch it curving away through the fields towards the next stop of Stęszew, and thence to Poznań. Neither of us spoke as the train chugged into the distance, leaving behind a long, lingering trail of white smoke and that evocative and unique smell of a hard-working steam locomotive.

THE STRANGLED GOOSE

On the Sunday, Simon and I had had our first run together with Ol49-59 on the 05:05 to Poznań. This run was notable only for the fact that, not for the first time, I had slight problems braking the train to a halt anywhere near the exact spot marked at the end of platforms. Stanisław, the driver, became so infuriated with my braking – which only made my performance even worse – that he exploded, *'Jeff, you* scheisse *driver!'* This made his partner, Marek, snigger uncontrollably and Simon smirk. I forgave Stanisław as he was quite obviously either under the weather or experiencing the aftereffects of a hangover – or possibly both. I have three photos of him in the cab at different times that day; in all of them he is fast asleep!

Following an afternoon siesta, I was given the opportunity to have an evening cab ride as far as Leszno, in the cab of Tr5-65. In contrast to the morning's Ol49, this loco was an ancient 2-8-0 that was almost ninety years old. We were to haul three four-wheeled coaches – known affectionately as 'Thunderboxes' – as far as Leszno, where water would be taken at the shed. An electric loco would then haul the consist from Leszno to Wrocław to take part in the filming of a period drama for Polish TV. The whole journey was videoed from the tender by a Swiss dentist named Alfred, who scarcely uttered a word throughout the whole trip. Our crew was Irek and 'Not-Mad' Andrzej.

The cab on this loco was much smaller than I was used to, the firebox a completely different shape and the tender much lower. Leaving Wolsztyn for Leszno, I opted to fire but Andrzej quickly became dissatisfied with my efforts and put more coal in the firebox himself. I asked if I could drive instead and Irek readily agreed. It was a strange sensation to roll gently through the countryside with no stops and little braking to do. In the brief diary which I kept every year, I noted: *'The loco made strange noises, like a goose being strangled'.* Leaving me at Leszno station, Irek drove on shed for water and for the waiting electric loco to be attached to the front of the train in order to pilot it to its destination. I watched it pass through the station in the fading light, still making its 'strangled goose' noises, before catching the 21:16 railcar for the hour-long journey to Wolsztyn. There, a stunning full moon greeted me and I was able to take some very atmospheric photographs.

MY BEST DRIVE EVER

I had an early morning start on the Friday to photograph the 05:05 and then returned to bed. The rest of the day might have paled into insignificance had it not been for another 'first' for myself, instigated by Howard with customary little warning. After travelling by 'Ice Cream Cart' to Leszno, and a mid-morning snack of Polish sausage, Simon and I travelled by *INTERCITY* First Class for a ridiculously low price to Poznań. Another change of trains, and a change of clothes en route, took us to Środa, where we joined Marek and Janus on the footplate of Ol49-7, which was hauling a short excursion from Poznań around the area.

Our job was to take it back to Poznań where it would become the 17:12 to Wolsztyn, which we would also drive and fire. Janus was the driver and he gave me my first full drive of the entire return Poznań-Wolsztyn run from start to finish. With a few helpful words of advice from Simon, and under the watchful eye of Janus – and there was no better tutor – the journey was an utter joy from beginning to end, ranking as one of the most precious memories of all my Experiences. Janus let me make all the starts and all the 'notching up' with not a single word of advice. Whenever it was time to brake for a station, Janus would give a calm nod towards the brake handle and leave me to it, no words were spoken. However, I didn't quite get all of the stops exactly on the mark; Janus was unperturbed by this, although he gave me marks out of ten for each stop. My record was 9 ½; the lowest score I will keep to myself! He must have felt very confident in my abilities as he even let me shunt the stock at our destination and drive our loco on shed. Simon and I stayed on to help dispose of the loco. From there we took Andrzej (who had mysteriously appeared) and Janus, to Bar Bliźniak for a pizza and a pint, or the Polish equivalent of a pint. Or was it two …?

LAST DAY

The next day was to be my last for 2009 so I got up early one final time to watch the 05:05 leaving in the rain. The weather was fit only for videoing, especially the spectacular slip as the loco passed the depot. From there, I walked in the early morning drizzle to the outskirts of the town, to the suburb of Karpicko, in order to photograph that Central European phenomenon of a stork on its rooftop nest. There were plenty to be seen from the cab on the Poznań line but I had never had chance to take much notice of them. Simon and I worked the 13:31 to Poznań and the 17:12 return. I fired all the way, both ways, and thoroughly enjoyed it. Simon drove every inch both ways, our crew again being Janus and Marek. The latter was quite content to let me fire my own way; Janus was a little more exacting and demanding of the kind of fire I maintained but in his usual charming and polite way.

Each and every year that I went to Wolsztyn was memorable in some way or another but 2009 was an extra special year. I was allowed to drive every inch of the return journey from Wolsztyn to Poznań, and was able, thanks to Simon, to see more from the lineside than in any other year. And I clocked up the grand total of 604 miles on the footplate!

A two-coach train from Poznań arrives at Wolsztyn behind Pt47-65

Not long after dawn, Ol49-7 shatters the peace of the woods near Ruchocice

One of Bob's little shunters and Ol49-7 inside Leszno depot

Leszno scrap line

11

SHEDBASHING – POLISH STYLE

LESZNO – A BUNKER'S PARADISE

We visited Leszno depot many times on our footplate turns, calling in there for servicing, turning, watering and, on one occasion, a running repair. We were, therefore, quite familiar with its layout and the types of loco we would find in and around the place. As we were 'on duty', and dressed accordingly, we had the run of the place whenever we wandered round it, and were never stopped or questioned. One of the challenges of a visit was to obtain a photograph of steam, diesel and electric locos all in the same frame.

The place had an irresistible attraction for us so we 'bunked' it several times when we were 'off duty', bringing back in a small way that thrill of the illicit, taking us back to teenage trainspotting days. The excitement of avoiding the shed foreman, combined with the fear of being caught (never mind the possibility of police involvement) could be relived in a way that was not really possible even during the post-steam era of British Rail. And we were, of course, in a land where our command of the language was, to all intents and purposes, non-existent. We were never challenged, thus denying us the opportunity to produce our trump card: a business card with the name 'Howard Jones MBE' emblazoned across a sunset photo of an Ol49 in full flight. I still have mine 'just in case' ... In fact, nobody in the place ever gave us a second glance and we never once saw anyone who looked like management.

If arriving in Leszno as mere passengers, we found that the pedestrian route to the shed took us out of the station, across one of the two main lines by means of a level crossing, then through an old goods yard, before reaching the depot itself. In a deep-rooted, instinctive hangover from our youth, we would try to find the least open way in. It was, therefore, a somewhat circuitous and lengthy route, which could have been shortened by simply walking off the end of one of the platforms, over the main lines on a boarded foot crossing, and straight into the depot yard. We didn't have the nerve to do this on the way to the shed but always returned by this short cut. No railway employee ever even looked at us, in spite of the crossing being in full view of the station's elevated signalling centre.

The depot itself is an old steam shed, which has never been seriously modernised. The whole building oozes atmosphere; the sooty smell forever associated with steam sheds is still all-pervading. Slightly diluted by diesel fumes, it permeates the very fabric of the building. The block-paved floor is a treacherous hotch-potch of uneven bricks and paving slabs, liberally coated in oil and sludge, waiting to catch out the unwary. Close your eyes and you can be transported back in time to the days of steam. The 'star attraction' in the shed was the decaying hulk of a Kriegslok, Ty2-1086, which had obviously once seen use as a stationary boiler. We didn't actually visit Leszno until our fourth 'Experience' and the derelict 2-10-0 – from which the loco buffers had been removed, but not the tender ones – came as a very pleasant surprise. The sunlight playing through the shed windows and the juxtaposition of a more modern electric loco made for some of the most atmospheric photos I ever took during our Polish adventures. By the time of our 2018 trip to Poland, the 'Kriegslok' had gone.

Parowozownia Leszno, as it is known, is still capable of heavy steam locomotive repairs, having a wheel drop which is in occasional use for Wolsztyn locos, as well as for diesels. Inside the shed in 2005, we found Ol49-7 having major repairs carried out on its cracked mainframes. At the time of our visit, the front pony truck had been removed, along with the leading driving wheels, leaving the loco as an 0-4-2. A spectacular shower of sparks beneath the loco came from a welder building up the sizeable crack with weld material.

The shed did not seem to have a large number of working staff but their ingenuity and versatility were to the fore one day in 2011, when a lubricator linkage broke on Ol49-7, as related elsewhere, in a case of 'all hands to the pump'. The same year, Ol49 59 was discovered to have a loose tyre on one of the rear driving wheels. The following day, it hauled the early morning train to Leszno, its place on the return journey being taken by a diesel, whilst Ol49-7 was steamed up at Wolsztyn to replace it. The next day, when visiting the shed whilst our loco was being serviced, we found it outside as a 2-4-2. Two days after that, it was inside the repair shop with the repaired and freshly painted offending wheel back in place. The very next day it was in steam, running up and down the shed yard. It probably hauled a train the day after, but it was the day of our return to England so we never found out.

Almost all of the roads in the shed have pits beneath them, another relic of steam days. The depot is divided broadly into three sections. Firstly, a two-section area which houses the wheel drop, an overhead crane and a set of screw jacks for lifting locos. Secondly, the main servicing/painting section which has seven roads, each with pits, and, finally, a section of three roads which defy definition. One of these roads contained the old Kriegslok; even by the time of our first visit, it looked as if it had been out of use for some years. Its rusting presence added a tangible atmosphere to that section of the shed, a potent reminder of the days of steam. The rest of this area often saw diesel or electric loco bogies having welding work done on them. One year, it contained the bare frames and smokebox saddle of an unidentifiable Ol49, which was undergoing a complete overhaul, its repaired boiler resting on a wagon in another part of the shed complex. In true Continental style, each road has a number above it. These numbers commenced at '10' because one part of the semi-roundhouse was derelict and in hazardous condition in the early years of our visits. It was demolished – or fell down – sometime around 2010.

At the station end, the shed yard contained row upon row of withdrawn diesel locos in various states of dereliction. Some had obviously been there a long time, judging by the plants and flowers growing in the cabs. The most common design of loco in the scrap line was the Romanian-built, Sulzer-powered ST43s. In our earlier years of visiting

Poland, we would see and hear these in action, occasionally in Wolsztyn itself. They shared the same diesel engine as the smaller Sulzer-engined locos of British Railways. To hear one was very nostalgic, taking us back to the Highland lines in Scotland or to freight operations in the north of England. The highlight of the scraplines was a Michelin-style single railcar, SN80-12, built in 1964 and which, according to some sources, is actually preserved, though you would not have known by looking at its peeling paintwork. Bizarrely, there is a very crude wooden model of this railcar at Wolsztyn depot.

Some years, we would find newly painted electric locos standing in the yard, waiting to be returned to service; in their blue 'PKP cargo' livery they looked very smart for their age. One year, SM42-295 was photographed inside the depot building in the same new colour scheme, but in 2007 EU07s were still being repainted in the PKP two-tone green livery, which suited them very well when fresh but quickly became flat and lifeless in service.

After a period of seven years during which I did not visit Poland, I returned in 2018 with Bob. One of our 'expeditions' was to Leszno, specifically to 'bash' the shed. In fact, we enjoyed it so much we did it again several days later! We thought at first that little had changed, though the sidings on the station side of the depot seemed to have gained even more withdrawn diesels, especially ST43s. Many examples of this class had been lying around for over a decade, so paint was flaking off all over, leaving great bare patches of rusty metal. Some of them had had their builders' plates removed, some sported the odd broken cab window. A couple of the locos that had broken windscreen glass had them boarded up, which seemed to me rather pointless. Remarkably for Poland, not one of the thirty or forty lying around the yards bore any graffiti; it seems that Polish street artists do not visit railway depots.

Other diesel classes were present but not in anywhere near the same numbers; SM32 and SM42 switcher type locos were the next most common, along with the odd diminutive SM03 0-4-0 shunter. In amongst the rotting hulks of the diesels were also to be found about a dozen equally rotting hulks of steam engines, predominantly Ol49s. One of these posed us a great puzzle: Ol49-69 was one of a few which had no tender and was even lacking its rear driving wheels, masquerading as a 2-4-2. Why the puzzle? There was another Ol49-69 back at Wolsztyn. In steam! We never did discover which one was the real one and which one was the impostor.

Most of the steam locomotives were rotting miserably away in a wasteland at the back of the sheds; some of them had platework like a sieve. There was one cab lying haphazardly on the floor and one whole tender, minus its frames, similarly abandoned close by. Nearer to the diesel depot itself, we found a full set of frames, complete with smokebox saddle, for an Ol49. In among the 'scrappers', we also found two Kriegsloks, behind one of which was what looked like a fully repaired boiler secured on a makeshift cradle, perched on top of an old coach bogie.

Inside the old steam shed in 2018, we found an SM42 'Fablok' which had just been repainted into the new Regional Railways orange livery. As we continued our explorations, we were surprised to find that this company had recently built a brand-new maintenance depot, on what had been waste ground to the rear of the main shed, complete with its own company-liveried shunting locomotive. Rows of refurbished EMUs, and some newer units too, stood outside this shiny new depot, which was partially fenced off. For once, we lacked the nerve to venture inside but, in the distance, not in a position for photography, was at least one Regional Railways liveried electric loco.

For Bob, though, the star of the show of any visit to Leszno was the diminutive 0-4-0 battery electric loco, which was employed in moving dead locos around the depot and yard. None of these tiny locos seemed to have any form of numerical identification but we later found out that they are designated as EL16. By 2018, many of these had been painted orange but when first observed they were in a colour to which the closest approximation would be sage green. Bob was so enamoured with this loco that he built a working electric-powered model of one in 'O' scale. And no visit with Bob to any depot which possessed one, would be complete without him posing in a cab doorway!

POZNAŃ: THE GHOST SHED

In spite of the fact that the city of Poznań is a much bigger, much busier and much more important place than Leszno, its depot seemed relatively insignificant by comparison. We ventured into its interior several times but there was surprisingly little to see and very few people around. It seemed that relatively little work went on there, although there was a huge traverser inside the spacious shed building. Besides a single SU45 diesel and an ET22 electric loco, there were just a few coaches under overhaul the first time we went. One of Bob's tiny battery-electric four-wheelers was on hand to do any necessary shunting. Like Leszno, the buildings had a rundown feel to them but here there was even more justification. There was a row or two of 'scrappers' – all diesel – and a half-roundhouse which was locked up and semi-derelict. Outside it was a turntable which was still in operation for those Wolsztyn locos which would fit. A water crane was also provided for steam visitors, which had their fires cleaned and were oiled in the shed yard.

Just across the main lines from the depot where our steam locos were serviced, was a very large factory with the name 'CEGIELSKI' prominently displayed on it. This building had, for over a century and a half, produced thousands of locomotives for Polish railways; it had also produced many other items of machinery, even weaponry.

In all the years of visiting this depot, never once did we see a loco arrive 'on shed', apart from our steam locos. There were usually several idle diesels standing around but only once did I ever photograph any electric locos 'on shed', probably because they require relatively little servicing and so are usually stabled in a siding near the station, awaiting their turn to take over another express train. The occasional diesel used for station pilot duties tends to stand around the station area too and, as at Leszno, the challenge was to capture a photograph with steam, diesel and electric all in the same shot.

FRANOWO: 'PROSZĘ JACEK SAHA?'

The reason why the Poznań depot was so insignificant didn't register with us until 2018, when Bob and I enjoyed a steam-hauled railtour around the area. Passing by Franowo yard, there were freight trains and locos in abundance; we realised why we had never actually seen a freight train in all our years of visiting Poznań station. So, when we returned to Poland in 2019, we asked if a visit to Franowo depot would be possible. Howard being Howard and Poland being Poland, this was easily arranged. Our instructions were simple and, in Howard's usual way, delightfully vague. *'You can't miss the place – it's huge. Just go into the yard by the main entrance – nobody will stop you. You will pass what looks like an old goods shed on your right. Go in the offices and ask for Jacek Saha, he will be expecting you at eleven o'clock.'* Simple?

With much help from our Satnav, we got to the yard in plenty of time and drove in. As Howard had foretold, there were no barriers and no security so we were able to drive all through the complex unchallenged. Franowo marshalling yard was quite simply massive and would have swallowed up the likes of Britain's Toton or Tinsley yards several times over. Following Howard's instructions, we soon came to what looked like an old goods shed with offices next to it, which fitted the bill perfectly. Except that it was on the left. Surely but a minor detail?

As I can speak at least a dozen words of Polish, I was 'elected' to enter the office and ask for the aforementioned Mr Saha. Timidly, I climbed a dingy staircase with magnolia-coloured paint peeling off the walls and knocked on the door, before entering a room containing half-a-dozen railwaymen, most of them at computer terminals. '*Proszę Jacek Saha?*' I asked, far more confidently than I felt. '*Angielski*', I added, pointing at myself just in case they miraculously thought I was Polish. Unsurprisingly, they answered in their native tongue but the gist of it was that, no, he definitely was not there, followed by directions in very basic English, which I barely understood. '*Dziękuję*', I muttered in thanks and exited, trying not to look as embarrassed as I felt.

Back in the car, we drove another half a mile through the yard until we came to some more offices, this time more modern looking, on the opposite side of a raft of sidings. A little hesitantly, we crossed the tracks to these offices where a similar performance ensued. Back to the car and another drive to a third set of offices. Much to my surprise, here I found someone who spoke more English than I did Polish (not difficult), had actually heard of Mr Saha and even gave us the man's phone number. He also gave us what he assured us were definitive directions to the maintenance depot. I chickened out of ringing the elusive man, knowing that if he could not speak much English there would be little point.

After a few more minutes' drive, we began to feel that we were on the right track when we caught sight of several rows of out-of-service locos. A couple of hundred yards further still down the road and there was our goal: Franowo depot. It had only taken half an hour to find the place! Immediately upon driving onto the premises, we noted a row of four EU07 electrics, all in PKP Cargo blue livery, parked at the side of a brick building, which looked as if it too might once have been a steam depot. Unlike Leszno, it had been extensively cleaned and modernised. In the depot yard to our right stood a mixed assortment of diesel traction, mostly medium powered Bo-Bos, together with a few electric locos.

THE MAN WHO LIKES TO SAY YES

With more than a little trepidation, we entered the building and *'Proszę Jacek Saha?'* was wheeled out for the last time to the first person we saw – a bearded, sandy-haired young man. Wearing a blue lab coat, and bearing a passing resemblance to Ed Sheeran, he smiled, *'Ah, you are the English? I am Lukasz.'* Shaking hands cordially with us, he apologised unnecessarily for his supposedly poor English. He asked us what we wanted to see first and led us through the shed, where we snatched a few photographs, before reaching a grassy area at the rear. Here, there were a couple of dozen more out-of-traffic locos, split fairly evenly between diesel and electric. There was, however, some overhead wiring in place and some freshly ballasted track, so perhaps some of the locos were in store rather than withdrawn; only one of them wore any graffiti. There were several of the large electric ET22s, mostly in washed-out two-tone green, but at least one in already fading 'modern' PKP Cargo blue. From the rear of the depot, we walked along the outside of the opposite side of the shed building, where there were four more blue EU07s in a row, parked next to what looked like the sandhouse.

Throughout our tour, Lukasz answered all of our questions as well as he could, in English that was far better than he gave himself credit for. He told us that he had learned English at school and seemed really pleased to have the opportunity to practise on three real, live Englishmen. He explained that one of the projects on which he was working was the conversion of more pairs of EU07s into twin units; this would involve decommissioning one cab from each loco and close-coupling them together semi-permanently. He estimated a timescale of three years for this project – which was not unprecedented in Poland – but what impressed us most about young Lukasz was the pride which he displayed in his work and in PKP. He clearly saw it as both a career and a job for life.

Next, we moved into the well-equipped main building, which was nowhere near full. Raised up on jacks, and minus its bogies, was ET22-944 of PKP Cargo Logisitics. Like most of the other locos we saw, it also carried its unique twelve digit UIC identity number of 91 51 3 150 254-3, not exactly a trainspotter-friendly system! We asked Lukasz if we could walk down into the illuminated pit which it straddled; he seemed surprised that we had even asked permission. Above our heads ran an overhead crane for lifting and removing items from locomotives. More locos were in the other end of the shed; one particular one, SM42-1604, was having traction motors changed, its old units sitting to the side of each of its weathered and work-stained bogies. On another road, SM42-1601 was parked between two sets of overhead walkways.

'Lukasz, can we climb up there for a photo, please?'

'Of course!'

Again, the surprised look at the request. After that, we stopped asking and just did what we wanted. Lukasz never raised the slightest objection and we decided that he was 'The Man Who Likes To Say Yes'! Turning a corner, we came upon the loco of Bob's dreams: an EL16 in PKP Cargo colours.

Bob was in the cab like a shot, examining the control desk and comparing notes with his model.

Outside in the yard, we found SM42-1602 slowly propelling its classmate SM42-922 into the depot. Lukasz explained the differences between the bonnet tops of the two locos, one of which had a very squared off appearance whilst the other was much more rounded. I was busy photographing proceedings but wish I had listened more carefully to his explanation which I think had something to do with the exhaust systems ...

Sensing that our time was almost up, we took one last stroll along the lines of locos in the yard at the front of the depot. These were waiting their turn for overhaul or repair inside the shed. At the end of one line of medium sized diesels we found ST44-1245, one of the massive, ungainly Co-Co diesels. These beasts are ugly but impressive in their own way, being Soviet-era locos, built in vast numbers in what is now Ukraine. I knew them as 'Ludmillas' but Lukasz called this one a 'Gagarin', a name which I felt had a better ring to it. And on that note, after explaining a little of the history of the 'Gagarin', Lukasz accompanied us back to the main gate where we posed with him in front of EU07-349, or if you prefer, 5 140 009-9 2151. Shaking hands warmly with him, we bade him goodbye in our best Polish. And that, we thought, was that.

THE MAN WHO LIKES TO SAY NO

As we drove away from the depot to retrace our steps through the yard, we were waved down by a rather agitated, almost aggressive, young German, obviously a railway enthusiast. At first, he seemed completely baffled by our Englishness but we quickly realised that his bewilderment was caused by our hire car's German registration plates! He spoke but a little English; we spoke but a little German. Gradually, we cottoned on to the fact that he was looking for some special electric shunting locos with pantographs mounted on their central cab roofs. He described them as EM10s, and seemed astounded that we did not have a clue what he was talking about. We had not seen any on our tour of the depot but we knew where there were more stored locos, as we had seen them on the way in. We offered him a lift to try to find them, but he reacted to our suggestion in a very startled manner and flatly refused to join us in the car.

After a few minutes' awkward and uncomfortable conversation, we drove back towards the way out of the yard but soon noticed that there were, indeed, a few locos answering the description of electric shunting locos in the line of stored ones. Turning the car round, we went back, told him what we just had seen and where to find the elusive shunters. Again, we offered him a lift. Again, our offer was met with obvious suspicion, followed by a point-blank refusal. We gave up trying to persuade him, turned round and drove off. It was only much later that a little research revealed that the EM10s were a small class of locos which were ultimately deemed to be a failure due to their unreliability. The four that we saw at Franowo yard represented the whole class; they had not worked for a decade and would never work again. And we had not even bothered to stop and photograph them. The Wonders of Hindsight!

Like a volcano erupting, Pm36-2 departs Wolsztyn with a train for Leszno. Passing some of the 'preserved' locos in the former goods yard, the train is about to pass the south signal box

Dawn departures-Ol49-7 leaves Wolsztyn for Poznań whilst an 'Ice Cream Cart' departs for Leszno

Inside Wolsztyn depot, Ok22-31 awaits its next job

'Have you ever driven a mainline diesel? Yes, I have!'

12

IT'S NOT ALL ABOUT STEAM

HAVE YOU EVER DRIVEN A MAINLINE ELECTRIC?

During our 2008 Experience, things were not going to plan. Problems with the steam locos saw Ol49-23 inside the depot with several men working on it and lots of important looking components removed, its coupling rods amongst them. The elderly 4-6-0 OK22-31 had been in use, not an everyday occurrence and a sign of things not being right. By the Tuesday, two SU45 diesels had arrived to take over the steam workings, leading to endless complaints from several others who were on the course with us about 'losing their throttle time'. Howard would have no truck with their complaints, explaining that the steam loco was an old piece of machinery and if it was broken, then it was broken. However, a 'consolation prize' was coming our way, following a conversation which went something like this:

> Howard: *'Sorry, lads, no steam tomorrow. Have you ever ridden on a mainline electric loco?'*
>
> Julian and I: *'Er, no ...'*
>
> Howard: *'Would you like to? You might even get to drive it ...'*

Julian and I: *'Mmmm. Yes please!'*

Ten minutes later....

Howard: *'Take these pieces of paper to the end of Platform 6 at Wrocław Station. At about 13:30, a train will come in headed by an electric Class 83 lookalike in Regional Railways colours. Present your papers to the attractive lady in PKP uniform who will be standing by the locomotive and off you go.'*

So, we followed Howard's instructions to the letter, travelling firstly from Wolsztyn to Leszno by Ice Cream Cart, the Polish equivalent of our 'Pacers'. From Leszno, an EP07 electric hauled us to Wrocław, encountering an hour's delay en route at Szewce owing to a derailed wagon in a lengthy coal train. Unlike what would surely happen in the UK, we were allowed to pass right alongside this stricken train, albeit 'wrong line' as this route had bi-directional signalling.

Arriving at Wrocław Główny station with its graceful overall roof, we spent a pleasant few hours watching the trains go by. As at Poznań, we marvelled at the number of loco-hauled trains and at the use of station pilot locos for breaking up and reorganising train formations. Our stay was further enlivened by the appearance of EP07-0151, which had been repainted into British Railways' electric blue livery, complete with aluminium lion-and-wheel crest, in an acknowledgement of the British heritage of early Polish electric locos.

Everything happened as Howard had said it would. The Polish lady certainly *was* attractive and scarcely gave our permits a second glance. Within less than a minute, we were climbing into the cab of EP07-1003 and introducing ourselves to another Andrzej – the driver of the Poznań train, whose ultimate destination was Olsztyn, in the north of the country. He seemed totally unfazed by our presence, in spite of having even less English than we had Polish. At 13:35 prompt, we left Wrocław at the head of Train 65105, a semi-fast PKP Przewozy Regionalne (Regional Railways) train, retracing our route to Leszno and then onwards to Poznań.

A few miles out of Wrocław, with 125 kilometres per hour (about 75 mph) showing on the speedometer, Andrzej rose from his seat, waved a surprised Julian to sit in it, and then proceeded to instruct him by sign language how to drive and, equally importantly, how to stop. He showed both of us a booklet which contained the line speed limits for our route, kilometre by kilometre, and pointed to line-side marker posts. It was clear that he expected whichever of us was driving to adhere to the limits and during the journey he made occasional minor corrections to the settings of the controls to make sure that we did so. When a station appeared in the distance, he merely said, '*Stop*' and pointed. We were each allowed to do a couple of station stops and starts. Throughout the whole experience, a smiling Andrzej remained completely calm – which is more than could be said for two over-excited Englishmen!

After about a quarter of an hour of Julian driving, Andrzej motioned us to swap seats so that I could drive. It felt very strange to have in front of me what looked like a large steering wheel, which served as the power handle. Everything passed in a blur of sheer excitement and pleasure and I must confess that I have forgotten most of the technicalities of driving the loco, though I do remember having to press the red Driver Vigilance button at frequent and regular intervals. Years of practice on Polish air-braked steam locos ensured that both of us electric novices were able to stop the train with gratifying precision (yes, even me!) which must have impressed Andrzej and given him some small measure of confidence in us. After some time, we passed again through Szewce, where the stricken coal train was still standing in the platform. By this time, a breakdown train formed of coaches painted in a livery reminiscent of our Blue Pullman trains, had arrived on the scene and was parked in a loop with an SM32 diesel at its head.

One abiding memory of our journey was the number of heart-stopping moments when pedestrians would amble across the line not far in front of us, scarcely glancing up at our speeding train. Just as on our steam runs, cars and lorries would cross in front of us at ungated crossings. Andrzej scarcely blinked. Seeing this phenomenon from the cab of a steam loco, as we were used to doing, isn't quite the same as being able to see the whites of people's eyes from the front of an electric loco!

Our loco had good acceleration but the ride was little, if any, smoother than on our steam locos. At speed it certainly bucked around and I was reminded of a late friend of mine telling me how he once had a cab ride in a Class 87 from Carlisle to Preston that was so rough, he almost feared for his life. The view from the very front of a train was most interesting; we passed far more goods trains than we would have done in the UK. After my quarter of an hour in the driving seat, Andrzej took control again. If he was relieved to do so, he didn't show it. Stopping at Leszno, he drove the train the rest of the route to Poznań. From the point where the line from Wolsztyn joined the main line, the journey was even more enjoyable than when seen along the boiler of a steam engine. At Poznań, we shook hands and took our leave of Andrzej with effusive thanks. We tried to offer him some złotys but he stubbornly refused to take them. We had experienced 91 miles in the cab of a real mainline electric loco at the head of a real express train full of real, paying passengers. Climbing reluctantly down from EP07-1003, we posed for photographs alongside 'our' loco, grins stretching from ear to ear!

Julian and I each have a souvenir of our cab ride – PKP mugs with a picture on them of EP07-1003, the very loco we drove and even in the same livery. A pure coincidence that the mugs should portray 'our' loco but they are treasured mementos. From the sublime to the ridiculous – we travelled back to Wolsztyn on another uncomfortable Ice Cream Cart ...

HAVE YOU EVER DRIVEN A MAINLINE DIESEL?

The day after our electric cab ride, we were due to drive and fire to Poznań with Ol49-69. However, the paperwork permitting the loco to work the train had not come through so it was confined to playing up and down the shed yard with a couple of disgruntled course participants in the cab. The repairs on Ol49-23 were still incomplete so a second conversation along similar lines to '*Have you ever driven a mainline electric, lads?*' took place, offering us instead the chance to ride in the cab of the diesel at the head of the late morning departure for Poznań. And, hopefully, to have a drive too.

This time, the cab ride seemed to have a less official air to it, no paperwork was involved and we had the feeling that some złotys had changed hands 'under the counter', Mad Andrzej being undoubtedly the chief fixer! We had to purchase return tickets at the station ticket office beforehand, increasing our suspicions, but we were not going to question anything! At a little more than £2 each, the cost was of little consequence. Arriving at the station at around a quarter past eleven, Andrzej introduced us to our driver and we soon found ourselves in the cab of SU45-247, an old and smoky green Co-Co, at the head of the 11:34 cross-country service from Wolsztyn to Poznań. Before long we would be taking turns to drive the loco.

With a very Eastern European/Soviet bloc look to them, and painted in two-tone green, the SU45 Co-Cos were built in the 1970s and weigh 102 tons for a power output of only around 1750b.h.p. Broadly speaking, they are in the same power and weight bracket as our own Class 37s, and, like some of them, are fitted with Electric Train Heat – not that we needed it on our warm autumn day. This class of diesel-electrics could be seen on both freight and passenger turns; we were once hauled from Frankfurt Oder to Rzepin by one, probably the smokiest diesel ride ever! The diesel engines that power them were built under licence from FIAT and many of the locos themselves were built by Cegielski in Poznań, our destination for the day. I never found out whether or not our steed was built there. Their days on PKP were numbered as new generations of DMU and EMU entered service.

The cab of our loco had a distinctly old-fashioned feel to it and was well worn; it was the first we had ever been in with curtains at the windows to close it off from the outside world. When you consider that one year, we did a steam driving (and firing) turn in 42-degree heat, it is not hard to see why they are fitted. A useful feature, especially with so many stops, was the rear-view mirrors attached to the cabsides, similar to those fitted to the UK's North American-built Class 66. The diesel power unit throbbed rhythmically behind us but was not over intrusive.

Our driver was cautiously welcoming but, just like Andrzej the previous day, had next to no English and took control as we left Wolsztyn. Our drive was interesting to say the least; just to be able to see the line and take it all in from the very front of the train, rather than looking along a boiler barrel, made a refreshing change. The cross-country route from Wolsztyn to Poznań is just over 50 miles in length and is mostly single track, with passing loops at many of the sixteen stations en route. The last few miles are 'under the wires.' Understandably perhaps, the driver was a little nervous with novices in command of his train and allowed us each just a couple of station starts and stops between Wolsztyn and Grodzisk.

Nearing the latter, our suspicions of the ride being unofficial proved well founded when, without ceremony, our driver opened the door in the bulkhead behind us and bundled us into the noisy, smelly and oily confines of the engine compartment! And there we had to stay until we had passed the prying eyes of the signalman in the tall box on the platform; we remained there until we had passed the second signal box, a matter of great disappointment as the attractive signal lady was on duty that day! Eventually we were summoned back into the cab by a tap on the window of the door through which we had been evicted.

As with our electric machine, power was applied by what looked more like a steering wheel than a throttle control. Similarly, there was the constant presence of the Driver Vigilance Device to attend to if our train were not to come to a grinding halt. Acceleration seemed somewhat lethargic but time was easily kept with our load of just two Regional Railways coaches. The steam locos that we had driven were devoid of such a luxury as a speedometer so it was interesting to note the kind of speeds which we attained, with a maximum, I noted, of about 85 kph – or 53 mph if you prefer. The highlight of our journey was descending the bank from Wiry and crossing onto the main Berlin-Warsaw line. The euphoria from this was short-lived as at Luboń, the first stop on the main line, and without warning, our driver opened the cab door and unceremoniously ejected us from his domain, indicating to us that from here on we were no longer welcome. We just made it into the first carriage and as the guard came to check our tickets, we knew why we had had to purchase them in the first place …

In just under two hours from leaving Wolsztyn, we arrived in Poznań station with plenty of time to pose for the customary photographs alongside 'our' loco before it ran round its coaches in readiness for its return journey. We didn't try to push our luck and ask for a cab ride back to Wolsztyn, even though we hung hopefully around near the SU45 for a while. The driver merely acknowledged our presence with a curt nod. Instead, we rode back 'on the cushions' on the 15:27. Maybe we should have offered him a few złotys?

PKP PRZEWOZY REGIONALNE spółka z o.o.

DOLNOŚLĄSKI ZAKŁAD PRZEWOZÓW REGIONALNYCH WE WROCŁAWIU

ul. Marszałka J.Piłsudskiego 105
50-085 Wrocław
tel.: (48-71) 717 55 00
fax: (48-71) 717 55 09
pr.wroclaw@pr.pkp.pl

Permit Nr /238/2008

for locomotive cabride while hauling regular train or for travelling in the train consist

Mr/~~Mrs~~ JEFF NICHOLLS who has produced passport No 807146201 as identification, is permitted to ride in the locomotive cab while hauling passenger train **No. 65105** between **Wrocław Gl. - Poznań Gl.** or in the train consist on the date **02.10.2008r** .

DYREKTOR
mgr An...rzej Piech
z ... Dyrektora
... ...owych

....

Manager, PKP Regional **Railways, Dolny Śląsk division in Wrocław**

PLEASE NOTE!

1. Before entering the locomotive cab, this permit is to be show n to the locomotive driver.
2. The person travelling in the cab is to obey the directions of the locomotive driver and to take place shown by the driver.
3. "PKP Przewozy Regionalne" spółka z o.o. / PKP Regional Railways Ltd. is not to be held responsible for any damage to the belongings or c10thing of the person travelling.
4. "PKP Przewozy Regionalne" spółka z o.o./ PKP Regional Railways Ltd. is not to be held responsible for any injuries sustained due to any possible event.
5. On arrival to the destination specified in this document, the permit is to be handed back to the locomotive driver.
6. When requested by the control body, the person travelling is to show a valid ticket for the specified route.
7. I hereby accept the above mentioned rules and declare to obey them .

'Have you ever driven a mainline electric? Yes, he has!'

Ol49-59

Ol49-69 and 'Thunderboxes'

13

EXCURSIONS AND EXHIBITIONS

2005 SPECIALS

In 2005, Bob and I, together with Julian and his wife, Renate, were fortunate to watch, crew or ride on the Schoolchildren's Specials run that year for local pupils during Children's Week. The specials operated on the short run from Wolsztyn to Tuchorza and back with Tr5-65. This deceptively powerful vintage loco, with the number '65' on its chimney in cast numerals, was entrusted to haul a mere two four-wheeled coaches. Bob was lucky enough to drive one of the trains, whilst I walked up to the site of the old Skansen station on the Zbąszynek line to photograph the special as it passed me. This was the week for special workings and on the Thursday evening I found no less than five locos in steam at the shed: Pt47-112, Ol49-69, Pm36-2, Ok22-31 and Tr5-65. This was the most locos I was ever to see all in steam at the same time.

At this time, there was also a series of so-called 'Mushroom Picker' specials, which used the vintage four-wheeled stock for hair-raising trips down the freight-only line to Konotop. On the Friday, I worked the 09:45 special with Ok22-31, driving one way and firing the other. Our Polish crew was

Piotr and Josef, the latter for only the second time during my visits, and my partner was Peter, a quiet but highly competent East Midlander. Unlike my freight trip on this line a few years previously, the Polish crew didn't insist on driving, nor did we lose any cinderguard glass, mainly because our dainty little loco was much narrower than the lumbering great Pt47 I had ridden on the timber train. I lost count of the number of times I almost lost my precious PKP cap though, as branch after branch swept the train during our passage of the 'jungle'. In days gone by, the train would stop from time to time for passengers to alight and pick mushrooms. On our trip nothing of this nature occurred; I suspect the train was more of a Boozers' Special than anything to do with fungi!

On returning tender-first to Wolsztyn, the Ok22 ran round its train and worked non-stop to Poznań with the same crew, carrying some very drunken revellers in the vintage coaches. Apart from the Polish crew, I was the only other person on the footplate for this leg and was allowed to drive the whole journey, bar the beginning and end. This was the first time I footplated this route without station stops, an absolute joy with no injector to work and next to no braking to worry about! At Poznań, after disembarking our merry passengers, we went on shed to service, water and turn the loco before returning empty stock, non-stop to Wolsztyn, again with me driving on all the cross-country sections. A long day on the footplate but most enjoyable.

JANUS'S VIRILITY TEST

In 2018, Bob and I went to Wolsztyn for a week for a variety of excursions and 'experiences', each of them unique in its own way. Much to our surprise, the first day we were there, Howard told us that we were to footplate Ol49-59 on the 11:27 to Leszno and back, me firing on the outward leg and Bob firing the return leg, but neither of us driving. A ticket would have to be purchased for the one not on the footplate, who would travel in the train. The reason for this unexpected bonus became clear when I climbed aboard at Wolsztyn station, begging a pair of gloves from Andrzej – not for the first time! In the cab was a couple, man and wife, both at least twice my weight and three times Bob's! It was obvious that neither of them would be fit, able or even willing enough to wield a shovel for any length of time.

Before all this took place, we had to pass Janus's daily depot virility test. At 9:30 most days there was a briefing/debriefing in the Dyspozytor's office of anyone connected with the Experience. Janus, who had been retired early from his beloved footplate, would make everyone a cup of coffee. All very hospitable, and he did serve very tasty cake to accompany it. But, as we had discovered years before, Polish coffee has the strength of industrial paint stripper. Janus made it in a smallish glass mug, in which he lovingly placed two-and-a-half large spoonsful of coffee grounds. After stirring it – no milk or sugar allowed – Janusz would watch you drink it ... We had the feeling that he knew exactly what he was doing; he would watch with a faint, knowing smile hiding behind his silvery-grey moustache as people tried to suppress their grimaces. If you were very lucky, he would be called away and there would be a mad dash to the sink to pour the stuff down. If you were unlucky, it was a case of 'grin and bear it'. Fortunately, Polish cake is delicious and helps to drown the taste.

Back on the 11:27, I fired nearly all the way to Leszno. At one point I offered the shovel to the apprentice, who struggled to put two shovelsful in the firebox before wiping her brow, sighing and wordlessly handing me back my implement. In the driving seat, her husband was panicking Andrzej as he seemed either not to listen to, or not to understand any instructions he was given. Part way through the run, the woman took over the driving whilst her husband slumped, exhausted, in the fireman's seat, refusing to even contemplate a turn on the shovel, let alone risk any attempt to operate the injector! His wife didn't seem to have a clue what was going on either and Andrzej had to spend the whole time that she was 'driving' guiding her every move, with his hand over hers, just as he had done with her husband.

Not far from our destination, Piotr, who was crewing with Andrzej, insisted on taking over from me and cleaning the fire, so he must not have been very impressed by my somewhat rusty efforts. Maybe he just felt sorry for me as there were no *curvas* involved … Bob, however, was allowed to fire all of the way back by Andrzej. That evening, we ate from a Turkish takeaway, the only eatery we could find open in Wolsztyn at eight in the evening!

BY 'THUNDERBOX' TO KIERKZ. WITH HAMBURGERS

When Bob and I travelled to Wolsztyn in 2018, it was the first time either of us had been back for at least seven years. It was a pleasure, therefore, to find that two of our old friends, in the shape of the ever-reliable Ol49-59, and classmate

Ol49-69, were still in action. I had not seen the latter in steam for ten years and, in the meantime, it had received a major overhaul and been bulled up. The smokebox door was now adorned with a 'Fablok' style maker's plate similar to those carried by some diesels. Just below this plate was a PKP plate similar to those carried on the cabsides of most steam locos (and to the one sold to me by Marek.) There was also an inverted horseshoe bolted to the front of the right-hand running plate; we never found out the significance of this adornment. The smoke deflectors had received special treatment – a white-painted design which, to our eyes, resembled a Native North American motif. We were unable to decide whether or not we liked it, but it was certainly very striking, making the loco instantly recognisable.

The day after our exploits assisting the struggling couple, we witnessed yet another first for us as Ol49-59 and Ol49-69 backed their empty stock trains side-by-side out of the station, a few hundred yards in the direction of Zbąszynek. Number 69 was at the head of the four-wheelers, and Number 59, which had only just returned to steam after a day's minor repairs (a boiler washout and a new brick arch) was in fine fettle at the head of the two blue-and-yellow Regional Railways coaches for the Leszno service. Once given the road, they whistled simultaneously, long and loud. With voluminous smoke, steam and sound, they charged downhill towards the station. From our vantage point on the footbridge by the level crossing at the north end of the station, they made a sensational spectacle as they ran in parallel.

Also in the area at this time was a party of American railfans, led by one who was particularly loud and bossy, sporting a very obvious blond wig, and who we nicknamed 'Mr Syrup' (from Cockney rhyming slang: *syrup of figs – wigs).* He was attempting to organise all the photographers on the bridge whilst shooting videos himself with a tripod-mounted camera. He was very loudly annoyed by anyone moving on the footbridge and making it vibrate, but his whole manner was such that it was very difficult not to defy him. Adding to his frustration, many local Poles used the footbridge to go about their daily business whilst the crossing gates were closed. Bob and I exchanged meaningful glances, rolled our eyes and sniggered quietly to ourselves. This was not to be our last encounter with Mr Syrup.

With the two trains now in their respective platforms, we made our way to the bridge over the south end of the station. Janus had lit some oily rags in the smokebox of the preserved 'Kriegslok', which stood in the unused bay platform, so that it was possible to take a photo of what appeared to be three locos in steam. The stage was now set for another Wolsztyn special: the simultaneous double departure. The train of four-wheelers behind Ol49-69 was timed to leave for Poznań and beyond at 11:27, exactly the same time as the service train for Leszno behind Ol49-59.

Three tracks leave Wolsztyn at the south end of the station; our trains would take the two outer roads. It was one of those times when you would dearly love to be in two – or even three – places at once, as photographers and videographers were allowed to stand in the centre road since that track was disused. In a nod to Health and Safety, they were supervised by a man in an orange reflective jacket. The crews of the two locos were in contact by mobile and the plan was for the Leszno service to pass under the Poznań line bridge at exactly the same time as the special train passed over it. And so it came to pass; everything went like clockwork. We ran side-by-side until the lines split, the Leszno train eased off and our loco was opened up to charge up the bank to the bridge, the two trains simultaneously crossing each other with military precision.

Our train of four-wheelers reached Grodzisk around noon. We were booked for a 50-minute lunch break stop and for our loco to take water, courtesy of the Fire Brigade. We adjourned to a nearby takeaway for hamburgers and chips, the limit of my Polish vocabulary. Leaving Grodzisk, we headed non-stop to Poznań, arriving there in our usual Platform 4b. Here, the loco ran round its train and we departed, tender-first, for a lengthy and convoluted tour of the Poznań area, taking so many turns and junctions that we quickly and completely lost our bearings! As we wound our way through the inner suburbs of Poznań, one of the first things to catch our eye was the huge goods yards at Franowo, where we noted a twin-unit, electric freight loco, amongst many others. We would return to the yard the following year for a proper visit.

Passing through the outer suburbs of the city, we came to a more agricultural area and our eventual destination of Kiekrz, which I had visited in 2009. The island platform here was, as at many smaller Polish stations, quite narrow. Ol49-69 uncoupled and ran round our train to begin the return by a different route, then giving us another first by steaming non-stop through Poznań station. We then followed another tortuous and circuitous route before re-entering the station facing the right way for our chimney-first return to Wolsztyn! At Grodzisk, during a short break and 'pit stop' on the return journey, the Fire Brigade again refilled the tender. As at all the breaks, we were glad of a rest and a stretch. Whilst the four wheeled coaches ran remarkably steadily, the wooden seats lacked any kind of padding, so our posteriors were in need of relief. And there were, of course, no toilets on board …

During the whole of the day out, pairs of tour participants were allowed to ride in the cab. From Grodzisk to Wolsztyn it was our turn to ride the footplate, but we did no driving or firing until we reached our destination, whereupon Bob was allowed to do some of the manoeuvring – that is until Andrzej jokingly announced his dissatisfaction with Bob's efforts and took over in a very theatrical manner. We left the train in the station and went back to our hotel for a wash and brush-up before dining at the appropriately railway-themed 'Lokomotiva' restaurant.

HOW DO YOU PRONOUNCE 'MIĘDZYRZECZ'?

The following day, an 08:26 start from Wolsztyn, behind a sparkling Ol49-69, reunited us with the 'Thunderboxes' for a gentle run as far as Zbąszynek, taking us along the Berlin-Warsaw main line once again. It was interesting to see things from a different perspective as we travelled on the open platform at the front of our coach. Several freights passed us, heading in the opposite direction, including a lengthy container train hauled by a red DB electric loco and a cement train with a twin electric unit at the head. Arriving at Zbąszynek, we were surprised to see, in the adjacent platform, a small two-car diesel unit wearing Arriva colours, the same livery as could then be seen on routes to and from Wales. We had noted one on the Poznań line the previous day. Similar in size to the well-loved (or hated) British 'Pacer' units, but with buffers, MD4280 proclaimed 'Międzyrzecz' ('between two rivers') on its destination blinds. Its crew jumped out to photograph our train on their mobiles, before it departed for the same destination as us.

Our loco ran round its train and set off tender-first, following the Polish Pacer and passing through Swiebodzin, before reaching Międzyrzecz, where Bob and I detrained. Here, our train was to become an excursion train for local people to enjoy the novelty of a short trip in ancient coaching stock behind an old steam locomotive. In an adjacent platform, SM30-507, a preserved 0-4-0 diesel shunter, stood attached to a short rake of three old four-wheeled coaches, similar to ours, giving rides along a short length of track into what was to become the scene of a small-scale Railway Exhibition. Our train was booked for a forty-minute break here for the loco to be watered and serviced, so there was time for a brief look at a few of the exhibits which had already arrived. We were able to snatch a few photos of them before climbing back on board our now-packed train for a round trip to the city of Gorzów. The station here was situated some way from the city itself in an area full of builders' yards, but sported a remarkable signal box, which resembled a fairy tale tower with two balconies, all painted in a pale green colour and roofed in bright red tiles.

After a leisurely run round, we departed chimney-first on our return journey, stopping at Skwierzyna, where the usual crowd of locals surrounded our train. In the opposite platform from us stood a long train of open bogie wagons headed by one of the lumbering 'Gagarin' diesels, its two-man crew both wearing fluorescent orange PKP gilets and one of them also wearing a matching fluorescent baseball cap. Intrigued by our steam loco, they descended from their cab to marvel at the sight, and to watch it run round its train. Shortly after one o'clock, we arrived back in Międzyrzecz, where several more exhibits had arrived for the show. Leaving our train, we walked through what looked like it had once been the station goods yard; everything had an olden-days atmosphere, there was little or nothing 'modern' about it. Just outside the station, a vintage orange-liveried 'Autosan' bus was running a service to the nearby village of Nietoperek via the town's main square. (Autosan is a long-lived Polish builder of buses.) The bus driver's name Krzysiek – Chris – was on display in the windscreen; the man himself proudly wore a uniform cap. In the yard was a small number of food stalls, so we treated ourselves to a hamburger and an ice lolly each. I even managed to ask for mayonnaise with the burger and salad. Just!

A fair sized crowd of people had turned up for the event, which was free. Lined up in the old goods yard on several parallel tracks were: the Arriva DMU we had seen at Zbąszynek, a modern 'LOTOS' Bombardier built Bo-Bo diesel with the unmemorable number of 3 650 023-6 3105, the preserved SM32 we had seen earlier and EU07-E238, an electric loco painted in 'Captrain' livery. The line-up was completed by the arrival of our steam loco, which had left its train of four wheelers in the station. The line alongside the goods bay had recently been relaid with concrete-sleepered track and fresh ballast, an encouraging sign, but the rest of the yard had a rundown, weed-infested appearance. On the opposite side of the main line, a modern centre-cabbed LOTOS loco shunted a long rake of rusting chemical tankers, its driver seemingly far more interested in the exhibition than in his own job. And who can blame him? Also on the opposite side was a characteristic octagonal water tower with a very Prussian air about it. On 'our' side of the tracks there stood a brick-built signal cabin, its signalman keeping an eye on proceedings. A few of the onlookers ventured onto the second track away from us, but the presence of two green jacket-clad 'Straz Kolej' deterred them from going any further. Although carrying batons, at no point did we see either of them remonstrate with any of the crowd; everyone just stayed back as the processions began.

Leading off was the blue and silver LOTOS diesel, which ambled up towards the station and then back to its original position. As soon as it stopped, and in a cloud of blue diesel smoke, the Arriva unit revved its engines and began to run towards the station in parallel with the SM32, which had uncoupled from its ancient four wheelers. We watched some of this from the goods bay and some from the cab of the electric loco, which was open to visitors, although there were no ladders or platforms to help with the climb from virtually track level. It was possible to walk right through this loco from cab to cab, which we did, but there was no railwayman in either cab to explain anything to inquisitive visitors, or to monitor their behaviour.

As we climbed down from the cab, we were followed by

a Polish family. The mother stopped us and asked in very good English what brought two Englishmen to her home town. It took some explaining, but she was thrilled to find that anyone from overseas would want to visit Poland. She was even more amazed that we had driven the steam loco that was in the yard – we let her think that we had driven it to Międzyrzecz – and that we had come from Wolsztyn. Like most Poles we ever spoke to, she was unaware of the significance of the place. (A decade or so earlier at Wolsztyn, a young Polish man had apologised to me for our steam locomotive and was at great pains to explain that this was not the modern, forward-looking Poland of which he was so obviously proud. Only then did I tell him that I had just driven it from Poznań to Wolsztyn and that the real reason for our visit to Poland was to see and ride on the 'old-fashioned' steam trains!)

Various combinations of exhibits now took their turn to parade before the enthusiastic crowd. The SM32 returned to its three coaches but didn't couple to them; instead, our Ol49 coupled onto the small diesel and double-headed it through the yard. The crew of the steam loco was Piotr and another man I didn't recognise. Piotr was absolutely in his element and wore the biggest smile I have ever seen on any Polish engineman. He waved royally to the crowd, acknowledging all who addressed him. His fellow crewman didn't look quite so enthusiastic and stood in the cab doorway yawning through some of the proceedings!

On the next track across, the LOTOS diesel coupled up to the Arriva unit and hauled it through the yard towards the station, before running back and propelling it almost to the end of the track it occupied. The unit then uncoupled, ran up to the signal box, joined the main line and was last seen disappearing towards Międzyrzecz station and beyond. The LOTOS diesel did one more runpast, then disappeared in the same direction as the Arriva unit. Finally, the preserved SM32 coupled back onto its four-wheelers, and, with a huge cloud of white smoke, headed vigorously for the bay platform where we had first seen it. Our Ol49 headed off to the station to rejoin our 'Thunderboxes', ready to depart for Wolsztyn. The show was over; it was time to make our way back to our train in time for our return.

We departed chimney-first at 15:30 and made a steady run as far as Skwierzyna, where we had stopped on the outward journey. Here, we paused so that our loco could make yet another run round, again with a crowd of interested locals milling about, both on the tracks and on the platform. In an adjacent siding, switched off, stood the 'Gagarin' which we had seen earlier in the day. Its wagons were now filled with timber and an SM42 diesel was coupled 'inside' it. Once our 2-6-2 had uncoupled and run down to the signal box, I decided that I would like a photo of the freight train, so I simply walked across the track and took the photos that I wanted. When in Rome! There was one more run-round to undertake, this time at Zbąszynek where we watched a Berlin-Warsaw express arrive behind a modern electric loco, similar in outline to the LOTOS diesel we had seen earlier. And, like that diesel, it didn't display any kind of class number, just its twelve-digit UIC number.

Arrival back in Wolsztyn at 18:26 was somewhat of an anti-climax but, again, a relief for our posteriors which had spent too many hours on the hard bench seats! After dinner at Lokomotiva, we returned to our hotel, the excellent Kaukaska. There, we discovered another phenomenon that was new to us: the Polish wedding ... It went on VERY noisily until well after five the next morning. About once an hour, the music would stop for some ten minutes, presumably while the band tanked up on more 'wodka'. Our brief slumbers were rudely interrupted as the band played a mixture of Polish songs and American Country and Western numbers. On and on. And on … At two o'clock I gave up and went for a walk round the deserted streets of the town. A night to remember but for all the wrong reasons!

MORE PARALLEL RUNNING

After our precious few hours of sleep, a somewhat complicated, but unprecedented, day followed our excursion to Międzyrzecz. To start proceedings, a tender-first Ol49-69 took passengers in two green bogie coaches as far as Rakoniewice, where it ran round them and returned to Wolsztyn. Here, Ol49-59 was already at the head of two more similar coaches and there was time to photograph the two specials in the station, although they were not side-by-side. At 10:32 we departed in Number 59's train as far as Stefanowo, where we stopped.

At Stefanowo station there was a tiny museum dedicated to Polish railways, supervised by a young Pole in his twenties. This lad was obviously more than a little eccentric, wearing a PKP uniform and trying to order everyone about. I recognised him as the young lad I had met at least once before, in the cab of an Ol49, when he had tried to tell me how to fire the loco. There was also a beautiful station sign with the letters of the station name fretted out on the front of a lamp, making a brilliant foreground for photography. Some ten minutes later, Ol49-69 arrived with its special. The two trains sat in parallel at the island platform, posing for photographs. Here, we had our second encounter with Mr Syrup, who tried in vain to organise the photographers into a line so that he could get his 'shots'. At one point, he became so frustrated that he said, and I would like to think that he was joking, '*We are Americans you know and some of us may have guns.*' This had the opposite effect from what he anticipated and people continued to wander at will around the station. It is just possible that *some of us* may have 'accidentally' strayed in front of his lens ...

After a wait of another ten minutes or so, Ol49-69 whistled loudly and its passengers reboarded their coaches before the loco made a vigorous departure and everyone remaining at the station managed to get their photographs – perhaps even Mr Syrup. Another ten-minute wait ensued before our train departed for the short run as far as Zbąszyń, the first station on the Berlin-Warsaw main line. This station has a large, quite grand building which appeared to be unused, the actual platforms being very stark and windswept. There was a feeling of a different, bygone era about the whole place; it felt like the setting for a Cold War spy film. The small town of Zbąszyń was notable as being the place where hundreds of German Jews fled in 1938. It was not too difficult to picture their arrival at this grim and bleak spot.

Ol49-69 had already uncoupled from its train and run round it, so that by the time of our arrival it was tender-first and ready for the next leg of our excursion. We climbed down from our train to photograph proceedings as Ol49-59 ran forward with its two coaches and backed them on to the other train, thus combining to make a four-coach train, top-and-tailed by our two locos. Led by Ol49-69, we set off along the electrified main line for our next destination, Nowy Tomyśl, a dozen or so miles away in the direction of Poznań. Here, we alighted to watch proceedings as the two trains disentangled and reorganised themselves, lining up with two chimney-first locos ready to engage in another parallel departure. This time, the two trains were to run side-by-side along the bi-directionally signalled main line back to Zbąszyń, and we would be on board one of them. We left the station in parallel, the guard of the opposite train leaning out of the front of the two carriages. In a sign of the times, he wore an orange hi-vis top but also sported a very stylish pair of sunglasses, even though it was a dull day. Two cameramen stood between the tracks filming and photographing our departure, one train on either side of them, yet another of those times when you would wish to be in two places at once!

AT MANY POINTS along the route, especially on station platforms, there were small crowds gathered to watch this amazing spectacle. But most remarkable of all, from the point of view of Health and Safety, on the footplate of Ol49-69 a man sat on the cab steps, one arm around the tender handrail and the other clutching a video camera, filming the entire run. This man turned up at many places we visited – he worked on the railways and was also a driver for LOTOS.

It is difficult to put into words the sheer excitement of

running alongside another steam-hauled express. I have never before, or since, been able to watch the workings of a steam engine at speed from such close quarters. To watch the valve gear in action, all of the rods moving rhythmically in unison to do their jobs, and simply to marvel at the whole spectacle, is something which I will never forget. Our two trains reached speeds that I would estimate as between 50 and 60 mph as they took it in turns to 'overtake' each other and then drop back to allow the other to pass. People even shook hands between the trains! Just about every head in each train hung out of a window, every passenger waving with the sheer exhilaration of it all. Cameras clicked away and mobiles videoed the spectacle. Each train must have overtaken the other at least three times during what was one of the most amazing and remarkable things I have ever seen on a railway anywhere in the world. I know it was not unique, it has been done in several other places, but to us it was an extra special experience which we will almost certainly never repeat.

All too soon, we were back in Zbąszyń, where we arrived in parallel. The footplate cameraman from Ol49-69 began to climb down the steps well before the train stopped, jumping down onto the track. Deftly, he uncoupled the loco from its train so that it could run round. Our loco too ran round its train but Number 69 departed first for Wolsztyn. It stopped again at Stefanowo for us to pass, the cameraman still clinging on like grim death, so that we arrived in Wolsztyn in time to reach the footbridge near the station and watch its arrival. It had been a day where there are just not enough superlatives to express the sheer excitement, drama and pleasure of what we had witnessed on a modern railway system. As for it happening in the UK? One can but dream!

In the evening, we were supposed to have a barbecue and fireworks at the shed but bad weather meant that, instead, we had a meal for all the tour participants in the Kaukaska Hotel. We spent much of the evening attempting to avoid constant shots of 'wodka' from Andrzej; those that we were unable to avoid went into a nearby pot plant! There was a melancholy mood around the room that evening as the wide-screen TV on the wall showed Poland playing in the 2018 Football World Cup. And being eliminated ...

Tr5-65 heads a Schoolchildren's Special past Skansen

Photographed from the cab of an electric loco, Ol49-69 is on display at the railway exhibition held at Międzyrzecz. Some of the other exhibits can be seen in the distance

Parallel running! Ol49-69 pounds along the Warsaw-Berlin main line alongside our train headed by Ol49-59

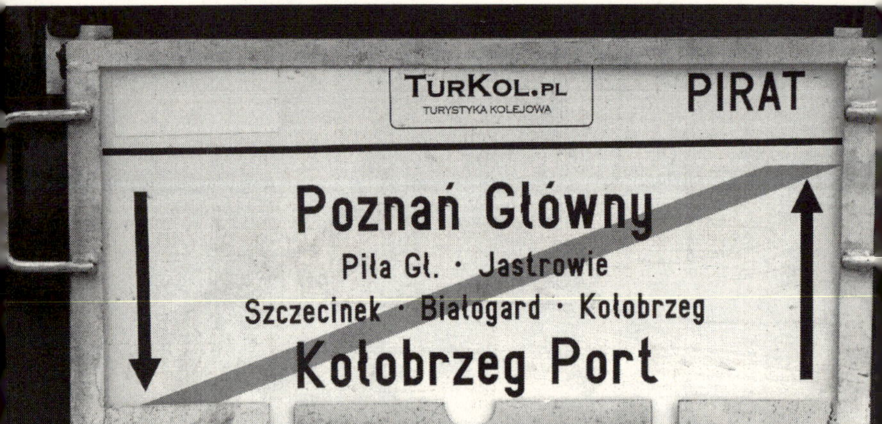

'Pirat'

14

AND FINALLY, 'PIRAT' – A VERY SPECIAL SPECIAL

WHAT TIME??!!??

Most years, a steam-hauled special train – 'Pirat' – runs from Wolsztyn to Kołobrzeg, on the Baltic Coast in the north of Poland. In 2019, three of us signed up to travel on it as part of a four-day trip to Poland, little knowing what a marathon it would be and that we would witness even more sights which would be impossible in most countries.

The evening prior to the trip, Howard gave us our tickets and a few details about the excursion. One minor detail he omitted was the departure time from Wolsztyn, so I casually put the question to him. His answer came with a completely straight face:

'Twenty-five past three'.

'Seriously? What time?'

'Seriously. Twenty-five past three!'

Arriving at Wolsztyn station with some twenty minutes to spare, we parked on the waste land next to the railway, to find that only two of us had brought our rucksacks. One of us had not … A mad dash in the hire car back to the hotel and there it was lying on the ground. I heaved a huge sigh of relief as it contained my passport, camera, clean underwear etc. ... My mercy mission was made easier by the fact that, late at night in Wolsztyn, most traffic lights switch over to flashing amber, so there is no frustrating wait for a red to clear to green on an empty road.

Back at the station with just five minutes to spare, and having regained my composure, there was just time to brace my camera on the top of a redundant bufferstop to photograph Pt47-65 at the head of our seven-coach train. The 2-8-2 had recently had a major overhaul and a coat of very dark green paint so it looked resplendent, simmering in the gloom of the station, its three large headlights ready to pierce the darkness of that unearthly hour. Most passengers, Howard included, had by now joined the train and ensconced themselves comfortably in compartments. Promptly at 03:25, and with no regard for the sleeping inhabitants of the town, the whistle sounded and we set off towards Poznań. Travelling through the dark, we didn't stop at any of the stations, not even Grodzisk. As our train was running at the beginning of June, dawn began to break as we passed through the outskirts of Poznań. Shortly afterwards, we arrived at Główny station where our Pt47 uncoupled, then ran back past us to the depot for watering. Whilst it was slaking its thirst, a modern LOTOS Bo-Bo diesel locomotive – BR 285-129 – attached itself to the rear of our seven-coach train. The reason for this didn't become clear until much later in the day. Its driver, Sebastian, was often to be seen around Wolsztyn, a keen railway enthusiast as well as a professional driver. The last time we had seen him was hanging from a footplate clutching a video camera during our parallel running the previous year.

The three of us climbed down from the train to watch proceedings whilst several overnight *INTERCITY* expresses passed through. Our coaches were each adorned with boards reading:

Poznań Główny – Piła Główny – Jastrowie-Szczecinek – Białogard – Kołobrzeg – Kołobrzeg Port

The boards also had the name of the tour organiser, *'TurKol. Pl,'* an abbreviation of Turystyka Kolejowa, and PIRAT, together with a diagonal blue stripe. With Pt47-65 back at the head of our train, we departed Poznań at 05:30 and headed out of the station, the loco's cylinder drains emitting a loud whooshing sound, accompanied by dense clouds of white steam. By now, several of the travellers who had started out at Wolsztyn were asleep; the three of us were not among them! The holidaymakers who had joined us at Poznań were distinctly livelier, many of them being excited children brandishing Pirate flags and carrying buckets and spades.

Our train turned northwards towards our first stop, some two hours later, at Piła, the largest city in the northern part of Greater Poland. Here, our loco was watered by the local Fire Brigade, as was to become the norm throughout this marathon. The crew left the cab to oil round their charge and to check that all was well. Crowds of locals surrounded the train and children were passed up into the cab of either the steam loco at the front, or the diesel at the rear. Nobody asked permission from the drivers but the crews were not in the least bit bothered by groups of children in the cabs with their attendant dangers, not the least being a red-hot fire in one of them! Once the driver whistled to announce our imminent departure, all the visitors left the cabs. On the diesel they were replaced by several passengers from the train – including Russ – who joined Sebastian in the cab as far as our next stop. Our 2-8-2 had a couple of very lucky Wolsztyn Experience participants in the cab, in addition to the normal crew. Theirs was to be a very long couple of days, sharing the driving and firing with the regular crew.

At the next stop – Jastrowie – the riders climbed out of the diesel cab and were immediately replaced by more excited children from the nearby town. Once more, there was a long stop whilst another Fire Brigade refilled our tender and our big green machine was besieged by locals of all ages, several of them posing happily on the track in front of our train! After the excited children had been evicted, several others, myself included, joined Sebastian in the cab of the diesel as we set off on the next leg of our journey, which was to take us as far as Białogard in Middle Pomerania. The Bombardier-built loco was very hi-tech, with a computer screen and readouts of just about everything the driver would need to know, not that Sebastian seemed to take much notice of them. He spent most of his time in conversation with the passengers in the cab, speaking to people on his mobile or organising photographs on his laptop. Occasionally, he would make fine adjustments to the power control but, as far as I could tell, the diesel was giving no assistance to the steam loco, merely hauling itself along. By the time we reached Białogard, I was not at all sorry to leave the cab. There were plenty more passengers eager to replace me and there is only so long you can be interested by the rear of the last coach of the train, even when one of your travelling companions appears at the window of the end door to pull faces at you …

Throughout the journey from Poznań onwards, there was a relaxed yet lively holiday atmosphere. A few family groups, mainly of railwaymen, had made the early start from Wolsztyn whilst many more families had joined us at Poznań for what to them was probably just a cheap day at the seaside, with steam haulage thrown in. The train had been about three-quarters empty between Wolsztyn and Poznań, apart from the three of us, Howard, 'Boomer' John and Lovatt, who we had met on our narrow-gauge adventure at Środa.

DOWN TO THE SEA

We pulled into Kołobrzeg station at 12:35. Large crowds awaited our arrival, some of them perched on the running plate of a shunting locomotive parked in the station yard. The rest of the spectators were scattered all around the station platforms and yard, all under the watchful eyes of two railway policemen and the station signalman. After a wait of about a quarter of an hour, there began another of the most remarkable railway journeys which I have ever made anywhere in the world.

Our performance began with a false start … Mysteriously, we travelled a short distance along an incorrect route, the points leading us the wrong way, in spite of showing that they were set the way we wanted. This could have been a major setback; not so here. After hasty braking, our train casually reversed the necessary few yards, the recalcitrant points were attacked with a crowbar (!) and we began our slow, amazing passage through the streets of Kołobrzeg, to the very end of the line in the port area, a mere hundred yards or so from the Baltic Sea.

Once on the correct route, our train proceeded, at no more than walking pace, out of the station yard, passing perilously close to the bumpers of parked cars. Crowds of people stood close to the track, many with pushchairs or babes-in-arms. Pedestrians crossed the road directly in front of our train whilst, with great gusto, the merry passengers waved their Pirate flags out of the carriage windows. There was an air of great excitement, a festive atmosphere, on the train. Some passengers opened carriage doors and stood or sat on the steps to watch proceedings, and to wave their flags.

Leaving the station yard, we crossed a main road under the supervision of a railway employee, who stopped the traffic. Not without some difficulty! Once the main road had been negotiated, we entered a wooded park – 'Park Linowy Kołobrzeg' – where a man stood BETWEEN THE RAILS filming us approaching him head on. Did anyone react? Andrzej gave a brief toot on the whistle and the cameraman moved off the tracks in plenty of time. During all of this, most of the coach doors of our train were open so people would hop off, take a photograph, and hop back on again. We suspected that a few of the onlookers 'stowed away' on our train too!

Emerging from the woods, we crossed another street – Zdrojowa – again stopping the traffic but this time without supervision; cars simply halted, their passengers in awe at the sight of a full-sized steam engine and its train crossing the street. The line took us along this street to our next challenge, a roundabout, which our train negotiated without incident, passing more or less straight through the middle. Cars gave way to us with only a little reluctance, their drivers aware that their metal box was no match for 180 tons of steam locomotive! Next, we rolled down Toworowa, a street full of shops and apartments. Our route was lined with people, some of whom were evidently expecting to see us but many of whom were openly astonished at such a sight passing down their main street. Roads and pavements were jammed with parked cars, some quite close to our train. Few people could resist waving at us!

It was here that we came upon our next challenge: a pair of points which, like the plain track, were sunk into the tarmac but, just like the earlier ones, were set for the wrong direction. Much debate between the train crew ensued as to what should be done, heads were metaphorically scratched and the trusty crowbar was brought out of the cab of the diesel. Sebastian was joined by an anxious footplate crew, who oversaw the operation to move the rusty points, which took several minutes. There then followed a short debate as to whether or not it was safe to continue, the result of which was never in doubt. We continued at snail's pace with Sebastian lying on the ground to one side of the train and another member of staff lying on the other. (Surprisingly for Poland, both men wore high-viz tops.) Each of them had a walkie-talkie, which was in contact with the crew in the cab, and as each pair of wheels safely negotiated the points, they gave the all clear to the footplate crew. In this manner, the whole train took some ten minutes to negotiate the turnout.

The diesel remained unattended on the rear of the train, its engine idling.

From this point on, it was slow but plain sailing right down to the quayside, passing a small market on our left-hand side and more bustling shops and cafes on the other. Throughout all of this, shoppers crossed and recrossed our path, some indifferent to our presence but the majority watching on in wonder, waving enthusiastically at us. Our arms grew tired with waving back!

Just over an hour and ten minutes after arriving in Kołobrzeg, we reached the end of the line, in the shadow of Latarnia Morska, a lighthouse built to commemorate the end of World War Two. On the quayside next to us was a replica of one of Columbus's ships, the 'Santa Maria', running tourist trips around the harbour. The 'Pirat' was immediately surrounded by crowds of holidaymakers, many keen to 'cab' our Pt47, but the three of us quickly left to photograph this historic event, climbing up to the area around the base of the lighthouse to capture the moment from higher ground. Many locals and holidaymakers watched events from the top of the lighthouse, whilst four women wearing pirate hats posed for photographs, directly in front of our loco.

After an hour of posing at the end of the line, with the crew taking it in turns to be photographed at the cab window – especially Piotr, who once more wore what was probably the widest smile in the whole of Poland – Pt47-65 sounded its whistle and the diesel blew its horn to warn of our departure. Rather than travel on the train, we elected to follow it on foot so that we could take more photographs of the whole ensemble reversing through the streets and back to the station yard. The Pt47 had its cylinder drains open, hissing steam at spectators, its crew content to be taken along for the ride and making the most of their brief spell of fame. The reason for the diesel accompanying us was now obvious.

Once the train had made it to the station, we went to check in at our hotel, then back into the port to explore some of the seafront, dining in a restaurant where the menu was in Polish and German. But not English. Selection of food was done mostly by pointing but I was still quite proud to show off my expanding Polish vocabulary – *'proszę rahunec'* – as I asked for the bill.

ENCORE! WITH DIESEL

The following morning, we went back into town for another wander round the sea front, then headed back to the station to look for our Pt47, which we found simmering in the sunshine, its crew fussing over their charge and looking a little the worse for wear. The front coach had been detached from the rest of the train and was for footplate and support crews to sleep in overnight. Doubtless they had enjoyed plenty of *piwo* or *wodka* … People wandered freely around the station yard, admiring our loco, chatting to the crew, posing for photographs and climbing up into the cab. Much to our surprise, we found the LOTOS diesel at the head of our coaches, about to repeat the previous day's exploits of travelling through the park, across roads and roundabouts, past shops and apartments and down to the quayside. This time it hauled the train on its own, so we were content just to follow on foot and photograph the whole ensemble as it repeated the previous day's snail's pace expedition to the end of the quay. Bizarrely, half of the cab of the diesel was filled with white balloons!

The journey to the seafront took longer than the previous day's descent because, at one point, the train came upon a parked 4x4, whose rear end overhung the tracks. After a long wait, during which the local Police attended, and just when they and the crew had decided to 'bounce' the vehicle out of the way, its owner returned, letting out a long string of oaths and not in the least bit apologetic about the delay he had caused. Eventually, the diesel pulled up at the end of the line and, whilst it did not attract as much attention as the previous day's steam loco, there was still a steady queue of people, young and old, wanting to 'cab' it. After its hour in the sunshine at the quayside, the blue-and-grey loco slowly and cautiously reversed its train back towards the station. This time, a man in a high-viz vest and equipped with a walkie talkie, walked in front of the rear coach to give instructions to Sebastian.

Once the train had returned to Kołobrzeg station, the diesel ran round it, moving to the rear, whilst our steam loco moved across from the station yard to take its rightful place at the front. There being no turning facilities at Kołobrzeg, it would undertake the first part of our return trip tender-first. For a couple of hours, it sat simmering in the bay platform at the head of our train whilst a succession of people, ourselves included, climbed up into the cab. We sat in the sunshine, savouring the leisurely activities of a quiet terminus, especially the once-familiar sight of a wheeltapper making his way along the coaches of a long-distance express, at the head of which stood an *INTERCITY* electric loco, its pantograph lowered. As so often, its destination was unpronounceable.

RETURN TO WOLSZTYN. WITH WATER!

By 17:15, most of the happy passengers had flooded into the station and taken their places on the train. At 17:35 prompt, the guard's whistle blew, both steam and diesel replied in kind, and we began our nine-hour journey back to Wolsztyn, where we were due to arrive in the middle of the night. It was a pleasant journey through rural Poland, which is pretty flat, has a lot of trees and is punctuated at regular intervals by towns with real factories. Many of these big, sprawling, steamy installations with lots of shiny pipework, appeared to be fairly new. As on our other journeys, we noted how rare it was to see farm animals in the fields, apart from a few horses and cows. There were plenty of wild deer, eagles, kites, cranes and storks to be seen and once we spotted a wild boar.

We travelled south for some three quarters of an hour as far as Białogard, where Pt47-65 was removed from the train to travel the short distance to a small maintenance depot just to the north of the station in order to be turned. The sun shone brightly, as it had through most of the weekend. The 2-8-2 ran through weed-infested tracks adjacent to the platforms, towards the nearby depot, most of the passengers leaving the train to watch. Once the loco had passed us, we thought of the photo opportunities to be had at the depot, so walked confidently across the running lines and along the 100 yards or so to the depot to watch and photograph the spectacle in the late evening sunshine. We were not the only ones to do so – another 'When in Rome' moment – and, as at everywhere else on our journey, swarms of people had turned up to watch the spectacle of a steam locomotive passing through their town.

With little room to spare, our loco inched its way onto the turntable. The actual turning operation took only a few minutes but as the engine moved off the turntable, a young Polish father climbed on to it, his young son on his shoulders, to watch the loco move away. Once it had passed us, we scampered back across the tracks to retake our seats. Some twenty minutes since uncoupling, our loco was back at the head of its train, ready to depart on the next leg of its marathon. Whilst we awaited departure time, two northbound expresses came into the station. The driver of one, unusually wearing faded denims and white T-shirt, stood in the doorway of his cab watching our steam loco. It was hard to tell whether he was envious or just plain curious; similar reactions had been noted over the years from drivers of more modern traction. Just before our next stop at Szczecinek, we passed some sidings alongside another modern looking factory festooned with miles of shiny pipework and chimneys. Here, we watched the uniformed driver of an electric loco at the head of a train of sheeted bogie wagons, videoing our passage on his mobile.

THE CHARGE OF THE FIRE BRIGADE

There was still a lively and congenial atmosphere on the train, though by this time many children – and a few adults – looked the worse for wear and some were fast asleep. Many adults, us included, just stood in the corridor, leaning on the handrails, watching the countryside pass by in the twilight. Russ's journey was much enlivened by one of the women on the train, an attractive guard based at Wolsztyn, who insisted on tickling him in the ribs each time she passed him in the corridor. He didn't protest! At many of the stations through which we raced, small crowds of people had gathered to watch our passage and to wave at us. But the best part of the return journey was yet to come …

After another couple of hours, during which we crossed back from West Pomerania into Greater Poland, we stopped for water in the town of Okonek, between our outward stops of Szczecinek and Jastrowie. We pulled up at another narrow island platform and scarcely had our train come to a halt than it was swamped by the biggest tidal wave yet of waiting adults and children, who swarmed all around and all over the loco. Standing on the track, a grandmother posed with her two grandsons, who wore football shirts of clubs from across Europe. Our visit seemed to be the highlight of the year for the locals, who turned out in force to watch the spectacle; a queue rapidly formed for a visit to the footplate. In less than a minute, children were being lifted up into the cab or onto the front of the loco, some confident, some a little apprehensive, as they posed for photographs; several adults too posed for the cameras. The diesel on the rear was steadfastly ignored.

Water was supplied by no less than four fire engines, all parked up on what had once been a goods platform. There was much jollity amongst the fire crews, who not only seemed to be having the time of their lives but must surely have been working voluntarily, as many of them wore shorts, flip flops and T-shirts bearing the single word 'Straż'. Meanwhile, the footplate crew climbed down to oil round and check over their machine, ignoring the mayhem all around them and leaving the two Wolsztyn Experience participants to guard the footplate. During the course of watering, several main line expresses thundered through, doing at least seventy miles an hour. They sounded their horns in good time, and people stepped calmly back from the platform edge.

Once watering was completed, Pt 47-65 whistled to announce departure time. Those who had left the train climbed aboard and, with a whistle from the guard and long answering whistles from our two locos, we steamed into the gathering gloom. As we departed the station, the four fire engines lined up side by side facing the track and switched on their sirens, blue lights and full headlights. They then gave us a long farewell of full 'Blues and Twos', to the tune of which, along with prolonged whistles from our locos and much waving and cheering from the crowd and passengers, we left town!

THE LAST LEG

An hour and a half later, we stopped again at Piła, where no water was taken but some time was spent in the station. Strolling along the platform, we noted many children fast asleep, lying across seats and covered with jackets or jumpers. We arrived at Poznań around midnight, the majority of the passengers leaving the train, many of the children still asleep in the arms of their parents. Our loco uncoupled and ran to the depot to take on its final tenderful of water. Whilst it was away being watered, the accompanying diesel uncoupled and departed into the night, Sebastian waving cheerfully out of the cab window. The balloons had disappeared from the cab, but our driver must have had a very long weekend in charge of his diesel!

By a quarter to one, we were ready to embark on the final leg of our marathon, a non-stop run through the night, back to base at Wolsztyn, finally arriving at twenty past two, almost forty-eight hours since our original departure. The atmosphere on the train was very subdued during the last part of our trip; somehow, we managed to stay awake, in spite of having to be up at eight o'clock the same morning to pay a last visit to Parowozownia. There, we would say our 'thank yous' and 'goodbyes', and make one last brave attempt at passing Janus's virility test, (as he was a 'Wolsztyn Virgin', Russ had to provide the cakes) before driving to Berlin for our flight home.

We would not have missed the Wolsztyn Experience for the world. Any of it. Would we do it all again? If only ...

Okonek: our penultimate water stop with the 'Pirat'. Our loco briefly becomes a celebrity whilst the local Fire Brigade fill the tender. (C.R. Hatt)

Pirates! (C.R.Hatt)

The late Howard Jones MBE (C.R. Hatt)

The Man Who Liked To Say Yes. Lukasz poses with the author and Bob at Franowo Depot (C.R. Hatt)

P.K.P.

ACKNOWLEDGEMENTS

THIS BOOK IS dedicated to the memory of Howard Jones MBE, founder of the Wolsztyn Experience, who died far too young in June 2023. Without everything that he achieved, not a single one of the experiences described in this book would have been possible. A man of persistence and ingenuity, who managed to persuade the people in authority on Polish railways to do the unthinkable, his legacy cannot be understated. Without his dedication and determination, steam in Poland would have died in the last century. Thanks to Howard, my friends and I were able to experience far more than just riding on steam locomotives, and to enjoy experiences which, even now, I can scarcely believe really happened.

Altogether, I went to Wolsztyn a dozen times, enjoying the 'Experience' on the footplate for ten of those times, enjoying other activities on two more occasions. Visiting so many times, I accumulated a scrapbook of wonderful memories, some hilarious, some a little scary, many simply incredible, but all unforgettable. I had unbelievable experiences in the cabs of just about every type of motive power, though steam was the prime attraction. I met people, both men and women (inevitably, the men outnumbered the women by a considerable margin) of many shapes, sizes and nationalities, who travelled long distances to share in the excitement and drama of the footplate, to enjoy a meal out in like-minded company and to compare notes over or a beer (or vodka) or two, reliving their experiences with friends old and new.

My 'adventures' could only have happened in the wonderful country of Poland, indeed, the working title for this book was 'Only in Poland'. Sadly, the opportunity to enjoy 'The Wolsztyn Experience' finished at the end of August 2023, and the website has been closed down. At the time of writing, a limited steam service to Leszno and, at weekends, to Poznań, lingers on. If you have enjoyed the 'Wolsztyn Experience', you could undoubtedly add your own stories. If you haven't, you now have a good idea of what you missed!

Inevitably, in a tale of this nature, a modicum of technical detail is involved but I have tried to keep this to a bare minimum, preferring to recount the wealth of experiences which I and my friends enjoyed so much.

For the majority of the footplate experiences, I was accompanied by Julian Maddocks-Born, a friend of over sixty years since our childhood in Wolverhampton. Four times, I went with my long-standing friend, Bob Bowden, and once, I went on my own. My last two visits were to travel on various railway excursions from Wolsztyn; Bob came along for both of those, accompanied by a mutual friend, Russ Hatt, for the final one. I am grateful to all of them for their company in sharing some of the most unbelievable railway experiences of our lives. Grateful thanks go to Bob for the huge effort he put into the map at the front of this book, and to Russ and Julian for the use of their photos. Joe Monty from Massachusetts kindly let me quote from some of his writings in the bi-monthly American publication 'Live Steam Magazine' and editor Steve Rabone published the chapter on driving diesel and electric trains in two issues of 'Traction' magazine, one in 2014 and one in 2016, although I have modified the chapter slightly compared to the original articles.

Grateful thanks to the two special ladies who have helped so much in producing this book: to my editor Lorna Brookes of Crumps Barn Studio, for having faith in my story; and to my wife, Carol, for supporting my endeavours, her meticulous checking of my writings, and for taking on the mountain of dirty laundry that has come back from Poland each time! And a special dedication to daughter-in-law Molly who, tragically, did not live long enough to read my tale.

All photographs were taken by myself unless otherwise acknowledged.

ABOUT THE AUTHOR

JEFF NICHOLLS is a former primary school teacher. He has had a lifelong interest in railways and model railways, with many articles published in the railway and model railway press. Accounts of his Polish diesel and electric locomotive driving experiences have featured in *Traction* magazine. He lives in Warrington with his wife.

If you loved this book, you'll love these other non-fiction titles ...

HEAVENLY TIDINGS & RAILWAY SIDINGS

Stuart Samuel

The warm and humorous story of a boy growing up in the Yorkshire town of Halifax in the fading days of the woollen mills. His first adventures are spent riding the railways in the driver's cab, and a lifelong career as a parish priest begins when the local vicar joins the family for an evening of television

ISBN 9781999870584

Colkirk Tales

Alfred Absolon

Alfred Absolon's memoir is a unique window into life in rural Norfolk before the Great War. Colkirk is a place where folklore is as real as the seasons and the harvest is gathered by men and horses. The threshing machine is powered by a steam engine, and the village is home to traditional craftsmen who practice a fading way of life.

This is an authentic and unforgettable first-person account of farm life in a Norfolk village at the turn of the 20th century

ISBN 9781915067159